Introduction to
Quantum Computers

Introduction to
Quantum Computers

Gennady P Berman
Gary D Doolen
Ronnie Mainieri
Theoretical Division and Center for Nonlinear Studies
Los Alamos National Laboratory

Vladimir I Tsifrinovich
Polytechnic University, New York

World Scientific
Singapore • New Jersey • London • Hong Kong

Published by

World Scientific Publishing Co. Pte. Ltd.

P O Box 128, Farrer Road, Singapore 912805

USA office: Suite 1B, 1060 Main Street, River Edge, NJ 07661

UK office: 57 Shelton Street, Covent Garden, London WC2H 9HE

Library of Congress Cataloging-in-Publication Data
Introduction to quantum computers / Gennady P. Berman . . . [et al.].
p.   cm.
Includes bibliographical references and index.
ISBN 9810234902 -- ISBN 9810235496 (pbk)
1. Quantum computers.   I. Berman, Gennady P., 1946–
QA76.889.I54   1998
004.1--dc21   98-23218
                                                                    CIP

British Library Cataloguing-in-Publication Data
A catalogue record for this book is available from the British Library.

First published 1998
Reprinted 1999

Cover design: The design of the quantum computer on the cover was conceived by the authors after reading in the note by Gary Taubes [31] about his quantum-computing coffee cup discussion with Seth Lloyd.

This book is printed on acid-free paper.

Printed in Singapore by Uto-Print

Preface

The field of quantum computation is rapidly evolving. Quantum computing promises to solve problems that are intractable on digital computers. Quantum algorithms can decrease the computational time for some problems by many orders of magnitude. The main advantage of quantum computation is the rapid parallel execution of logic operations achieved by using superposition (entangled) states. To build a working quantum computer several problems must be solved, including the utilization of entangled states, the creation of quantum data bases and implementation of quantum computation algorithms.

The book explains how quantum computation works and how it can do many amazing things. It is intended to be useful for students and scientists who are interested in quantum computation but face difficulties in reading the original papers and reviews.

In the Introduction we present a very short history of quantum computation. The basic ideas on the Turing Machine are explained in Chapter 2. In Chapter 3 we describe the binary system and Boolean algebra, which are widely used in computer science. Some initial ideas on quantum computing are presented in Chapter 4. Using simple examples, we discuss the following quantum algorithms in Chapters 5 and 6: the discrete Fourier transform and Shor's algorithm on prime factorization. In Chapters 7, 8, and 9 we give an overview of digital logic gates and discuss reversible and irreversible logic gates, and how to implement these gates in semiconductor devices and transistors. Some important quantum logic gates are discussed in Chapters 10–14. A summary of unitary transformations and elements of quantum dynamics are given in Chapter 15. Quantum dynamics at finite temperature is discussed in Chapter 16. The implementation of quantum computation in real physical systems is considered in Chapter 17. In Chapters 18 and 19, we describe a realization of quantum logic gates in an ion trap. In Chapters 20, 21, and 22, quantum logic gates and quantum computation are discussed in linear chains of nuclear spins. Experimental logic gates and their achievements and possibilities are described in Chapter 23. One of the simplest schemes for error correction is discussed in Chapter 24. The dynamics of quantum CONTROL-NOT gate is described in Chapter 25. Quantum logic gates in a spin ensemble at room temperature are discussed in Chapters 26, 27 and 28. Concluding remarks are given in Chapter 29.

This is a many-author book, and each of us has contributed to different parts of the book. Berman, Tsifrinovich, and Doolen produced the first draft of the book. They were then joined by Mainieri who also produced the figures and tables for the book. In the rapidly changing field of quantum computation it is difficult to judge what should be covered in an introductory text, and we hope that we have covered the essentials.

We thank D. K. Ferry, L. M. Folan, R. Laflamme, and D. K. Campbell for useful discussions, and R. B. Kassman and R. W. Macek for critical reading of the manuscript. We thank G. V. López with whom the results on the dynamics of quantum logic gates were obtained. This work was partially supported by the Linkage Grant 93-1602 from the NATO Special Programme Panel on Nanotechnology, by the Defense Advanced Research Projects Agency, and by the Department of Energy through the Center for Nonlinear Studies and the Theoretical Division of the Los Alamos National Laboratory.

March 1998

G. P. Berman,
G. D. Doolen,
R. Mainieri,
V. I. Tsifrinovich

Contents

Chapter 1

Introduction

At present there are two basic directions on the intersection of modern physics, computer science, and material science. The first is the traditional approach, struggling to squeeze more devices on a computer chip. This direction is a central focus of nanotechnology – a modern science which uses a nanometer scale (10^{-9} m) to measure the size of electronic devices. Since the late 1980s, researchers around the globe have tried to create single-electron devices to replace the conventional MOSFETs (metal-oxide-semiconductor-field-effect-transistor). These devices operate by moving a single electron in and out of a conducting region. Single-electron devices may serve as transistors, memory cells, or building blocks for logic gates [1]-[7]. The single-electron transistor has evolved so that it is now possible, at room temperature, by applying a voltage to the operating electrode (gate), to transfer a single electron from a reservoir into a semiconductor island (so-called "quantum dot") surrounded by non-conducting material. Once an electron is in the dot, it blocks the transfer of other electrons due to the strong Coulomb repulsion (Coulomb blockade effect) [5, 6]. The current through a transistor depends on the number of electrons stored in the dot, allowing one to "write" and to "erase" the information. Another promising idea explores the use of molecules as naturally occurring nanometer-scale structures to design molecular devices [5],[8]-[11]. Devices in these classes take advantage of the quantum physics that dominates the nano-meter scale. All these devices are described by conventional current-voltage characteristics and are intended for traditional digital computers that operate using two values of a bit, "0" and "1".

The second approach is quantum computation, the main topic of this book. A quantum computer is intended not for accelerating digital computation using quantum effects, but to utilize new quantum algorithms which are not possible in a digital computer. In a quantum computer, the information is loaded as a "string" of quantum bits – "qubits". A qubit is a quantum object, for example, an atom (an ion) which can occupy different quantum states. Two of these states are used to store digital information. An atom in the ground state corresponds to the value "0" of the qubit. The same atom in the excited state corresponds to the value "1" of this qubit. So far, there is nothing new in comparison with the traditional digital computer except a higher density of digital information.

The main advantage of the quantum computer is not connected with the density of qubits. The difference is that quantum physics allows one to operate with a superposition of quantum states. For one atom, one can produce an infinite number of superpositional states using just two basic quantum states, which correspond to "0" and "1". For example, if two states have the energies, E_0 and E_1, one can prepare a superposition of states, "0" and "1", which corresponds to any average value of energy between the values E_0 and E_1. However, measuring the energy of a single atom, one can get only one of two results, E_0 or E_1, i.e., the states "0" or "1". To measure the average value of energy, one must use large number of identically prepared atoms.

Utilization of superpositional states allows one to work with quantum states which simultaneously represent many different numbers. This is called "quantum parallelism". What is the main advantage of quantum parallelism? If one has an efficient algorithm for calculation, like an algorithm for calculation of a sum, a product, or a power, then the superposition of numbers is not important. But there are problems which are considered today as intractable – problems which do not have an efficient algorithm. One such very important problem is the factorization of an integer. It can take thousands and thousands years for the most powerful digital computers to find the prime factors of a 200-digit number. A quantum computer can operate simultaneously on many numbers, leaving for an "observer" only the few desired numbers. The undesired numbers are removed by destructive interference. The usual comparison for this process is reflection of a light beam from a mirror. The reflected light is a superposition of photons moving in many different directions. Only one direction is selected by nature – the direction which corresponds to the law of reflection. Quantum computing makes use of a similar effect – constructive interference in the "desired" direction and destructive interference in all others.

Note, that unlike a digital bit which, in the process of calculation, assumes a definite sequence of values, "0" and "1", a qubit can be involved in a complex superposition of states with other qubits. One cannot determine the value of a specific qubit until the end of the calculation when the final measurement destroys the superposition. The output of quantum computation is very similar to the output of digital computation. The output is the same sequence of data obtained by measuring the state of the qubits: "there is voltage" (represented by "1"), and "there is no voltage" (represented by "0"). For example, after the action of the appropriate electromagnetic pulse, the excited metastable state of the ion produces a fluorescence which can be transformed into an electric signal. For the same input, one can get different outputs which correspond to the output from probabilistic digital computation. For more sophisticated schemes of quantum computation, for example, computation with an ensemble of nuclear spins at room temperature, an output is an electromagnetic signal (the signal due to nuclear precession) which can be analyzed by standard electromagnetic methods.

The history of quantum computing began with the academic question concerning the minimum amount of heat produced in one computational step. In 1961, Landauer showed that the only logical operations which require dissipation of energy are irreversible ones [12]. This led Bennet to the discovery of the possibility of reversible dissipationless computation [13]. Then, Toffoli suggested the famous reversible CONTROL-NOT gate (or CN-gate), which changes the value of a target bit ($0 \rightarrow 1$, or $1 \rightarrow 0$) if the control bit has a value 1 [14]. Toffoli also showed that reversible three-bit-gates (CONTROL-CONTROL-NOT, or TOFFOLI-gates) are universal for digital computation, i.e. combinations of these gates can produce any digital computation.

In the early 1980s, the idea of the quantum computer was introduced by Benioff [15] and Feynman [16]. They showed that bits represented by quantum-mechanical states can evolve under the action of quantum-mechanical operators to provide reversible computation. In 1989, Deutsch introduced the universal three-qubit quantum logic gate [17]. He showed that due to the exploration of a superposition of quantum states, quantum computation can be much more powerful than digital ones. In 1993, Lloyd proposed the implementation of quantum computation using electromagnetic pulses which induce resonant transitions in a chain of weakly interacting atoms [18].

In 1994, an explosion of interest in quantum computation was caused by Shor's discovery of the first quantum algorithm which can provide fast factorization of integers [19]. Shor's algorithm requires a time proportional to L^2 for a factorization of a number with L digits, compared with $\sim \exp(L^{1/3})$, for the best known digital computer algorithms. Quantum computers represent a potential threat to modern cryptography which assumes that fast factorization algorithms do not exist. In 1995, Barenco et al. [20] showed that a two-qubit CONTROL-NOT gate, in combination with a one-qubit rotation, are universal for quantum computation. This discovery made a quantum CONTROL-NOT gate of central importance for quantum computation. In the same year, Cirac and Zoller [21] suggested the practical implementation of quantum computation using laser manipulations of cold trapped ions. The first two-qubit quantum logic gate was demonstrated experimentally by Monroe et al. [22], who used the Cirac-Zoller scheme for a single Be^+ ion in an ion trap. Results on very interesting the Los Alamos trapped ion quantum computer experiment can be found in [23]. Turchette et al. demonstrated two-qubit quantum logic gates for polarized photons in a quantum electrodynamic cavity [24]. In 1995, Shor suggested the first scheme for a quantum error correction code [25]. His work stimulated a large number of papers which discuss different approaches to this problem. In 1996 Grover [26] (see also [27]) developed a fast quantum algorithm for pattern recognition or data mining. For N elements in the data base only about $\sqrt{N}$ trials are required for Grover's algorithm to find a given element, compared with $N/2$ trials for the classical algorithm.

In 1996, Gershenfeld, Chuang and Lloyd [28, 29], and, simultaneously, Cory, Fahmy and Havel [30] showed the possibility of quantum computation in an ensemble of quantum systems at room temperature. The experimental implementation of this idea (which utilizes a system of weakly interacting nuclear spins in molecules of liquid) is now being attempted [30]-[32]. One might think, that room temperature is incompatible with the idea of quantum computation, which relies on manipulation with complicated superpositional states. (These "entangled states" cannot be represented as the product of states of individual atoms.) Indeed, the interaction with the environment quickly destroys superpositional states. These superpositional states do not "survive" in our "classical" world. This phenomenon of losing quantum coherence is commonly called "decoherence" [33, 34]. Decoherence has a characteristic time-scale. Quantum computation must be done on a time-scale less than the time of decoherence. This is true for both the "pure" quantum system, at zero temperature, and for a room temperature ensemble of quantum systems (molecules). The characteristic time of decoherence depends not only on temperature, but also on the system. For nuclear spins, this decoherence time is long enough, even at room temperature. The main problem which prevented an implementation of quantum computation using room temperature ensembles is the following: How can one prepare a sub-ensemble, where only one state, for example, the ground state, will be populated? This problem was solved in references [28, 29, 30].

Discussions of a potentially realizable quantum computer involve a new field of investigation, quantum computer material science. This new field requires finding a medium which has a long enough characteristic time of decoherence. The theory of decoherence is the theory of relaxation processes for complex quantum states. Future development of this theory could significantly influence the progress in quantum computation. However, the problems connected with decoherence have no a direct relation to the main ideas of quantum computation. So we will not discuss decoherence further in this book.

The increasing number of reviews on quantum computation (see, for example, references [35]-[43]) reflects the rapidly growing interest in the field. At the same time, many students and scientists interested in quantum computation face all the difficulties common to any research which requires a knowledge of several different disciplines. A computer scientist often is not familiar with the ideas or even the terminology of quantum physics. Physicists have a similar problem with computer science. Overcoming this language barrier is the main reason for writing this introduction to quantum computers. The second reason is connected with our own work in the area of dynamics of quantum logic gates and quantum computation. This book includes the basic physics and computer science information necessary to understand quantum computation and the main directions in this quickly developing field. We avoid rigorous proofs and concentrate on specific illustrations which clarify the main ideas. At the same time, for simple examples, we present all necessary calculations. The reader can see how an idea works without omitting the details which often prevent the essential understanding of the whole idea.

We discuss almost all of the main topics of quantum computation which have been discussed in the literature. We consider Shor's algorithm and the discrete Fourier transform; quantum-mechanical operators (quantum logic gates) which are used in quantum calculations; physical implementations of quantum logic gates in ion traps and in spin chains, including an analysis of an ensemble of four-spin molecules at room temperature. We also discuss one of the simplest schemes for quantum error correction; correction of errors caused by imperfect resonant pulses; and correction of errors caused by the non-resonant action of a pulse. Because of the central importance of the quantum CONTROL-NOT gate for quantum computation, we included in this book our results on numerical simulations of dynamical behavior of this gate. We also present a short review of some basic elements of computer science, including the Turing machine, Boolean algebra, and logic gates. These are topics familiar to students of computer science, but are not well-known to many physicists. We also explain, where we felt it was necessary, the basic principles of quantum mechanics, which are probably not known to many computer scientists.

Our introduction is intended to be useful for students and scientists who are interested in quantum computation but do not have time or inclination to examine the original articles and reviews. We hope that this book will help a new generation of researchers who want to be involved in this new field of science which is expected to become of great practical importance. We also expect that this book will provide a new and deeper appreciation of the fundamental quantum phenomena.

Chapter 2

The Turing Machine

The simplest "theoretical" digital computer is the Turing machine [44, 45]. Here the word "digital" indicates that the computer operates only with definite numbers (and does not use any quantum mechanical superposition of states). This machine was suggested by the British mathematician, A.M. Turing. The Turing machine has three parts, a tape divided into the squares, a scanner, and a dial, as in Fig. 2.1. This machine can write a symbol X or 1 in a blank square, and erase them. Any positive integer is written as a sequence of 1's. For example, the number 5 corresponds to the sequence 11111. The symbol X indicates where a number begins or ends. For example, Fig. 2.1 shows two numbers 1 which are "prepared" for addition. The program for addition is presented in Tbl. 2.1. The symbol D is the command to "write the digit 1" in the corresponding square on the tape; X means "write X"; E means "erase"; R means "move the tape one square to the right"; L means "move tape one square to the left". The numbers 1 to 6 after the letter indicate the command to "change the dial setting to this number". The question mark represents a "mistake"; an exclamation mark means "job is completed".

Now we shall describe the process of addition. First, the scanner sees the number 1 on the tape, and the dial setting 1. The instruction on the intersection (1,1) is R1: "move the tape one square to the right, and

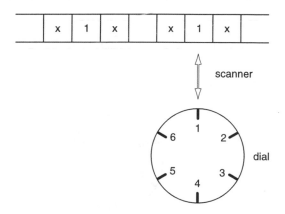

Figure 2.1: The Turing machine

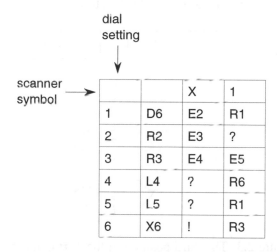

Table 2.1: The program for addition in the Turing machine.

set the dial to 1". The second position is shown in Fig. 2.2. Here, the

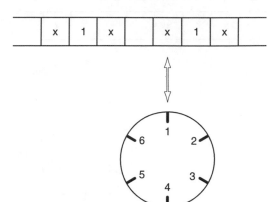

Figure 2.2: The second position of the Turing machine

scanner sees an X on the tape, and the dial setting is 1. The second instruction (1,X) is E2: "erase X, and set the dial to 2". The third position is shown in Fig. 2.3. The third instruction (2,□) is R2. Tbl. 2.2

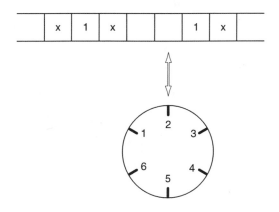

Figure 2.3: The third position of the Turing machine.

shows the sequence of positions and instructions following Fig. 2.3. The number in parentheses inside the square indicates the dial setting at the instructions

x	1	x	(2)		1	x	R2
x	1	x(2)			1	x	E3
x	1	(3)			1	x	R3
x	1(3)				1	x	E5
x	(5)				1	x	L5
x		(5)			1	x	L5
x			(5)		1	x	L5
x				(5)	1	x	L5
x					1(5)	x	R1
x				(1)	1	x	D6
x				1(6)	1	x	R3
x			(3)	1	1	x	R3
x		(3)		1	1	x	R3
x	(3)			1	1	x	R3
x(3)				1	1	x	E4
(4)				1	1	x	L4
	(4)			1	1	x	L4
		(4)		1	1	x	L4
			(4)	1	1	x	L4
				1(4)	1	x	R6
			(6)	1	1	x	X6
			x(6)	1	1	x	!

Table 2.2: The sequence of positions and instructions following Fig. 2.3.

position of the scanner. For example, 1(5) indicates that the scanner points to the square whose index is 1 and the dial setting is 5. If the scanner points the blank square, and the dial setting is 6, the corresponding notation in the Tbl. 2.2 is (6). The last row in the Tbl. 2.2 shows the result of addition: 1+1=2. The program for multiplication requires 15 numbers on the dial, but the idea of the programming is the same.

The Turing machine has the same main components that any computer has. The writing and erasing elements represent the arithmetic unit, which perform calculations. The table of instructions (Tbl. 2.2) is the control unit. The tape and the dial are the memory unit.

Chapter 3

Binary System and Boolean Algebra

Most "practical" computers make use of the binary system. In this system, any integer N is represented in the form,

$$N = \sum_n a_n 2^n,$$

where a_n takes the values, 0 or 1. For example, $59 = (111011)$, is a notation for,

$$59 = 1 \cdot 2^5 + 1 \cdot 2^4 + 1 \cdot 2^3 + 0 \cdot 2^4 + 1 \cdot 2^1 + 1 \cdot 2^0.$$

Let us assume that a practical computer will add the two numbers, 2 and 3. Because $2 = 1 \cdot 2^1 + 0 \cdot 2^0$, and $3 = 1 \cdot 2^1 + 1 \cdot 2^0$, we have in the binary system, the two numbers, (10) and (11). First, we add 0 and 1 (right column) to get 1. Then, we add 1 and 1 (second column from the right), and get 0 for the second column, and a carry-over of 1 for the third column. So, the sum is equal to (101). In the decimal system (101) is $1 \cdot 2^2 + 0 \cdot 2^1 + 1 \cdot 2^0 = 5$. A table for the addition of the binary digits (bits) is given in Tbl. 3.1.

In Tbl. 3.1, A is the value of the bit in any column of the first number; B is the value in the same column of the second number; C is the

13

A	B	C	S	D
1	1	1	1	1
1	1	0	0	1
1	0	1	0	1
1	0	0	1	0
0	1	1	0	1
0	1	0	1	0
0	0	1	1	0
0	0	0	0	0

Table 3.1: The table of addition in the binary system.

carry-over from the addition in the column to the right; S is the value of the bit in the sum, and D is the value of the carry-over to the next column to the left.

To work with this table, it is convenient to use the methods of the Boolean algebra [45]. These methods are especially useful, because, as we discuss below, the expressions written in terms of Boolean algebra are convenient for implementation in electrical circuits. A two-valued Boolean algebra can be defined by the tables of addition (Tbl. 3.2a) and multiplication (Tbl. 3.2b). In Boolean terminology the two operations are often referred as the OR and AND operations, respectively. The digits in the first row and column of each of the tables 3.2 refer to the values of each of the two input bits upon which the operation is performed, while those in the interior of the tables 3.2 give the value of the resulting output bit.

In terms of the Boolean algebra, the expression for S in Tbl. 3.1 can be written as,

$$S = \overline{(\bar{A}B + A\bar{B})C} + (\bar{A}B + A\bar{B})\bar{C}, \qquad (3.1)$$

where "bar" means "complement". (The complement of 0 is 1, the com-

(a)

	1	0
1	1	1
0	1	0

(b)

	1	0
1	1	0
0	0	0

Table 3.2: The tables of addition, (a), and multiplication, (b), for the two-valued Boolean algebra.

plement of 1 is 0). Let us check, for example, the second row in Tbl. 3.1. We have,

$$A = 1, \quad B = 1, \quad C = 0.$$

So,

$$\bar{A} = 0, \quad \bar{B} = 0, \quad \bar{C} = 1.$$

According to the Tbl. 3.2b,

$$\bar{A}B = 0 \cdot 1 = 0, \quad A\bar{B} = 1 \cdot 0 = 0.$$

Then, according to the Tbl. 3.2a,

$$\bar{A}B + A\bar{B} = 0 + 0 = 0, \quad \overline{\bar{A}B + A\bar{B}} = \bar{0} = 1,$$

$$\overline{(\bar{A}B + A\bar{B})}C = 1 \cdot 0 = 0.$$

The second term in expression (3.1) is equal to $0 \cdot 1 = 0$. So, the final value of the right side in (3.1) is, $0 + 0 = 0$, which is equal to the value of S in the second row in the Tbl. 3.1. The expression for D can be written as,

$$D = (\bar{A}B + A\bar{B})C + AB. \tag{3.2}$$

For example, for the second row in Fig. 3.1 we have,

$$(\bar{A}B + A\bar{B})C + AB = 0 \cdot 0 + 1 \cdot 1 = 0 + 1 = 1,$$

which is equal to the value of D in this row.

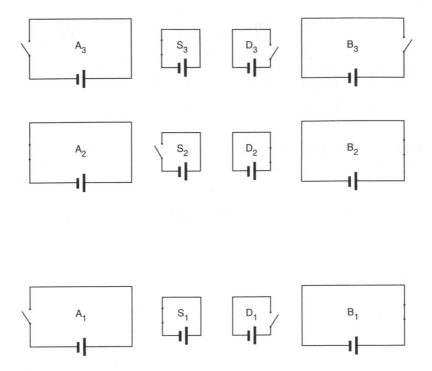

Figure 3.1: The left system of circuits, $A_3 A_2 A_1$, is loaded with the number 2. The right system of circuits, $B_3 B_2 B_1$, is loaded with the number 3.

Now, we can ask what is the simplest "practical" computer for addition, using Boolean algebra. Consider a system of circuits, each circuit having two current states – "current" or "no current". The first state corresponds to the value of the binary unit 1 and the second state corresponds to 0. We can write any number in the binary system using this system of circuits. Another system of circuits keeps the second number. In Fig. 3.1, the left system of circuits (A) is loaded with the number 2 ((10), in the binary system). In Fig. 3.1, the number 2 is represented in the form, $A_3 A_2 A_1 = 010$. The right system of circuits (B) is loaded with the number 3 ((11), in the binary system; $B_3 B_2 B_1 = 011$). The value 1 of a bit corresponds to the closed position of a switch (presence of a current in a circuit). The value 0 of a bit corresponds to the open position of a switch (no current in a circuit). Between the left and the right circuits one installs the systems of circuits, S and D, which hold the information about the sum and the carry-over, correspondingly. The main problems are the following: How does one operate on the values of bits A_1 and B_1 to obtain the values of S_1 and D_1? How does one operate on the values of A_2, B_2 and D_1 to obtain the values of S_2 and D_2? And so on. To do this, requires transformations (logic gates) which operate according to formulas (3.1) and (3.2).

These gates can be designed using a system of circuits. Assume, that we have three bits, A, B, and C, which are implemented by three circuits, A, B, and C, correspondingly. In Fig. 3.2, we demonstrate, as an example, the gate which transforms the values of the bits, A, B and C, into the value $(\bar{A}B + A\bar{B})C$ (the first term in (3.2)). In Fig. 3.2, we suppose that the switches "a" and "b" are closed if there is no current in the adjacent coils. If there is a current in the adjacent coil, the magnetic field of the coil forces the switch to be open. Special springs keep the switches "c", "d", "e", "f", "g", "h" open if there is no current in their adjacent coils. A current in the coils causes the adjacent switches to be closed. Assume, for example, that the value of the bit A is 1. So, there is a current in the circuit A (see Fig. 3.2). In this case, the switch "a" is open, and no current flows through this switch. If there is no current in the circuit A, there is a current through switch "a". This means that the current through switch "a" corresponds to the value of the complement A, $\bar{A}$. Correspondingly, the current through the switch "b" corresponds to the value $\bar{B}$.

Next, we have current through the switches "e" and "f" only if there is the current in the circuit B, and the switch "a" is closed. This means that the current through the switches "e" and "f" corresponds to the value, $\bar{A}B$ ($\bar{A}B = 1$ if $\bar{A} = 1$ and $B = 1$. Otherwise, $\bar{A}B = 0$). Analogously, the current through the switches "c" and "d" corresponds to the value $A\bar{B}$. The switch "g" is closed if there is current in the adjacent coil, i.e. if there is at least one current through the switches

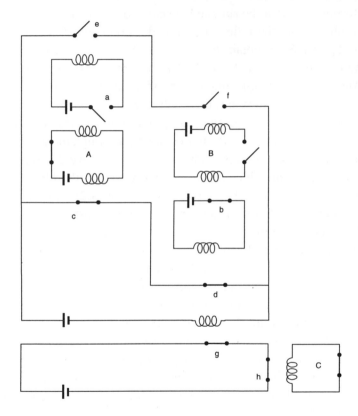

Figure 3.2: The logic gate which transforms the values of the bits A, B, and C into the value $(\bar{A}B + A\bar{B})C$, indicated by the current through the switches "gh".

"ef" or "dc". That means that the switch "g" is closed if at least one of the products $\bar{A}B$ or $A\bar{B}$ is equal to 1. So, the switch "g" is closed when $\bar{A}B + A\bar{B} = 1$, and open when $\bar{A}B + A\bar{B} = 0$. This arrangement thus implements the Boolean addition or OR operation. The switch "h" is closed if there is the current in the circuit C, i.e. C=1. This case is shown in Fig. 3.2. So, we have a current through the switches "g" and "h" only if $\bar{A}B + A\bar{B} = 1$, and C=1. It means that the current through the switches "gh" corresponds to the value of $(\bar{A}B + A\bar{B})C$. Analogously, using a more complicated scheme, we can arrange circuits with currents corresponding to the values S (3.1) and D (3.2). In modern computers, complex circuits are built using tiny silicon transistors, but the main idea of logic gates which transform the values of bits, is the same.

Chapter 4

The Quantum Computer

In a "practical" digital computer information is coded as a string of bits. In "quantum" computers, the elements that carry the information are the quantum states. For example, one can use two quantum states of an atom – the ground state and the excited state. The quantum system can be populated either in the ground state $|0\rangle$, or in the excited state $|1\rangle$. One might think that a quantum computer provides only an opportunity for greatly increasing of the density of bits. The reality, however, is much more powerful. A quantum system can be populated not only in the ground state or the excited state $|0\rangle$ or $|1\rangle$, but in any linear combination (or superposition) of these two states. That is why instead of the term "bit", the new term "qubit" (quantum bit) was introduced. The main advantage of quantum computation is that it allows one to make use of the technique of quantum parallelism, which can produce quantum computations that are even more powerful than massively parallel classical ones.

One can wonder how to use a superposition of qubits for deterministic calculations. Indeed, for the deterministic calculations considered above, we only can use the ground state and the excited state of the quantum system. So, in this case there is no distinction between bits and qubits. New opportunities arise in quantum computing because the computation do not have to be deterministic. Sometimes, it is more convenient to allow a computer to execute its steps randomly. This kind of computation can be called a probabilistic computation [39]. Usually, there are many different ways to arrive at the final answer, and each way has its own probability. If the probability of the very quick ways is high enough, the answer is found quickly most of the time. Then, probabilistic calculations can be used instead of deterministic ones. For example, there exists a fast algorithm for addition which we used for deterministic computation. But there are no fast algorithms for factoring. To find the factors of a number, we can try sequentially all natural numbers starting from 2 (deterministic way), or we can try numbers randomly with some restrictions (This is the probabilstic way).

If we use a superposition of quantum states, the computation will be also probabilistic, but different from classical probabilistic computation. There will be many different possible ways for a quantum system to attain the final state (final answer), but every way can be described not by the probability, but by the amplitude of the probability. Probability amplitudes are complex numbers, sum of which can add to zero (or cancel each other). The quantum computer will be efficient if only the correct answer survives with high probability, and the incorrect answers cancel each other.

Below we discuss Shor's quantum algorithm of efficient computation, following mainly the review of Ekkert and Loza [39]. The efficiency of computation is connected with the time of computation as a function of the size of the input. An algorithm is efficient if the time taken for computation increases no faster than a polynomial function of the size of the input. For example, the number N requires approximately $L = \log_2 N$ bits. (With L bits one can load any number from 0 to $2^L - 1$.) If there exists an efficient algorithm which computes the factors of N, it must have the number of computational steps S less than or equal to a polynomial function of L. It is known that any composite number N has a factor in the range $(1, \sqrt{N})$. If we try each number in this range to find a factor of N, it requires at least $S = \sqrt{N} = 2^{L/2}$ steps. The function $S(L)$ depends exponentially on L. So, this deterministic algorithm is not efficient. A quantum computer will have an advantage in comparison with a digital computer if the quantum algorithm is efficient for a problem which does not have an efficient digital algorithm.

The first efficient quantum algorithm was invented by Shor [19], for finding the period of a periodic function. Below we shall describe this quantum algorithm using a simple example of a periodic function, $f(x)$, where x takes only the integer values, $0, 1, 2, \ldots$. The problem consists of finding the period of the function $f(x)$ using Shor's algorithm. Assume that we have two strings of qubits. The string of qubits which holds the values of argument, x, we shall call the X string (register). The string of qubits which holds the values of the function, $f(x)$, we shall call the Y string (register). Consider, for example, a function $f(x) = \cos(\pi x) + 1$, with the period, $T = 2$. If the argument x takes the value 5, the value of function is $f(5) = 0$. These values of x and $f(x)$ correspond to the following states of two registers, X and Y, written in the binary system using the Dirac notation,

$$X: \quad |000...101\rangle; \quad Y: \quad |000...000\rangle.$$

Below we shall use the following notation for representation of the states of registers X and Y: $|x, f(x)\rangle$. For the case considered above we have,

$$|x, f(x)\rangle = |000...101, 000...000\rangle,$$

or in decimal notation,

$$|x, f(x)\rangle = |5, 0\rangle.$$

In what follows, we shall use more complicated states, $|k, f(n)\rangle$, and their superpositions, $\sum_{k,n} c_{k,n}|k, f(n)\rangle$.

According to Shor's algorithm, the register X is placed initially in the uniform superposition of all digital states. For example, if the register X consists of three qubits, the uniform superposition of $2^3 = 8$ digital states is,

$$X: \frac{1}{\sqrt{8}} \left(|000\rangle + |100\rangle + |010\rangle + |001\rangle + \right. \tag{4.1}$$

$$|011\rangle + |101\rangle + |110\rangle + |111\rangle) \; .$$

(One does not have to know the values of the function $f(x)$ in advance. These values are computed in parallel by a quantum computer [19]. There exist a standard digital algorithm that computes the function $f(x)$ for any value of x, see reference [39]. This digital algorithm can be realized with reversible digital gates. These gates can then be replaced by quantum logic gates, which can then be decomposed into a collection of two qubit CONTROL-NOT gates and one qubit rotations. Quantum logic gates act on superposition of the states $|x, 0\rangle$, and produce a superposition of the states $|x, f(x)\rangle$. A concrete example is given in Chapter 17.) In decimal notation, this is a superposition which can be written as,

$$X: \quad \frac{1}{\sqrt{8}}(|0\rangle + |1\rangle + |2\rangle + |3\rangle + |4\rangle + |5\rangle + |6\rangle + |7\rangle). \quad (4.2)$$

As one can see from (4.1) and (4.2), already at the first stage of computation, the quantum mechanical approach allows one to use a "superposition of numbers", which is impossible for a digital computer. The register Y, as before, holds the ground state $|0\rangle$ for all qubits. Next, the whole system XY of two registers is placed into a uniform superposition of states,

$$\Psi = \frac{1}{\sqrt{8}} \sum_x |x, f(x)\rangle.$$

In decimal notation it will be the following superposition,

$$\Psi = \frac{1}{\sqrt{8}}(|0, f(0)\rangle + |1, f(1)\rangle + |2, f(2)\rangle + |3, f(3)\rangle + \quad (4.3)$$

$$|4, f(4)\rangle + |5, f(5)\rangle + |6, f(6)\rangle + |7, f(7)\rangle).$$

The vector-diagram in Fig. 4.1 represents the superpositional state (4.3). Actually, the function Ψ in (4.3) is the wave function of the atomic system which represents registers X and Y. Next, according to Shor's algorithm, the register X transforms by the following rule,

$$|x\rangle \Rightarrow \frac{1}{\sqrt{8}} \sum_{k=0}^{7} e^{2\pi i kx/8}|k\rangle, \quad (4.4)$$

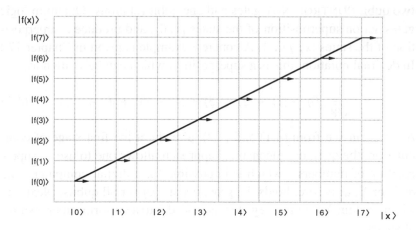

Figure 4.1: The vector-diagram for the superpositional state (4.3). Every vector on the intersection, $|x\rangle$ and $|f(x)\rangle$ represents the corresponding amplitude at the term $|x, f(x)\rangle$ in (4.3). The length of the vector is proportional to the modulus of the complex amplitude ($1/\sqrt{8}$, in this case). The angle between the direction of the vector and the horizontal line is the phase, φ, of the complex amplitude (0, in this case).

where x and k are written in decimal notation. For example, the state $|5\rangle$ transforms into the following superpositional state,

$$|5\rangle \Rightarrow \frac{1}{\sqrt{8}}(|0\rangle + e^{10\pi i/8}|1\rangle + e^{20\pi i/8}|2\rangle + \cdots + e^{70\pi i/8}|7\rangle).$$

The transformation (4.4) is a discrete Fourier transform for the register X. Now we apply this discrete Fourier transform (4.4) to the wave function Ψ in (4.3). As a result, we obtain the following new wave function,

$$\Psi' = \frac{1}{8} \sum_{x,k=0}^{7} e^{2\pi i kx/8}|k, f(x)\rangle = \qquad (4.5)$$

$$\frac{1}{8}|0\rangle\{|f(0)\rangle + |f(1)\rangle + f(2)\rangle + \cdots + |f(7)\rangle\}+$$

$$\frac{1}{8}|1\rangle\{|f(0)\rangle + e^{2\pi i/8}|f(1)\rangle + e^{2\pi i2/8}|f(2)\rangle + \cdots + e^{2\pi i7/8}|f(7)\rangle\}+$$

$$\frac{1}{8}|2\rangle\{|f(0)\rangle + e^{4\pi i/8}|f(1)\rangle + e^{4\pi i2/8}|f(2)\rangle + \cdots + e^{4\pi i7/8}|f(7)\rangle\}+$$

$$\cdots$$

$$\frac{1}{8}|7\rangle\{|f(0)\rangle + e^{14\pi i/8}|f(1)\rangle + e^{14\pi i2/8}|f(2)\rangle + \cdots + e^{14\pi i7/8}|f(7)\rangle\},$$

where $|0\rangle|f(0)\rangle$ means $|0, f(0)\rangle$; $|0\rangle|f(1)\rangle$ means $|0, f(1)\rangle$, and so on. The wave function Ψ' is represented by the vector-diagram in Fig. 4.2. The wave function (4.5) describes the entangled (mixed) state of the system of atoms (ions) corresponding to the qubits involved in the X and Y registers after the discrete Fourier transform of the register X. According to Shor's algorithm, one can find the period of the function $f(x)$ by measuring the state of the register X. Later we will explain how one can implement this wave function in physical quantum-mechanical systems.

Assume, for example, that the function $f(x)$ has the period $T = 2$, i.e. $f(0) = f(2) = f(4) = f(6)$, and $f(1) = f(3) = f(5) = f(7)$. In this case, we can rewrite formula (4.5) as,

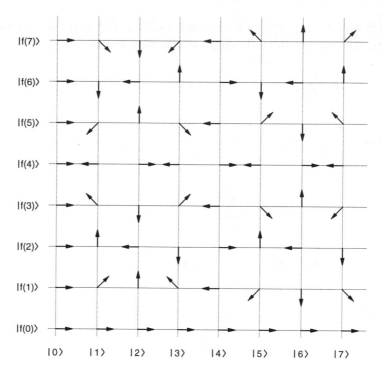

Figure 4.2: The vector-diagram for the wave function (4.5). The length of each vector is 1/8.

$$\Psi' = \frac{1}{2}|0\rangle\{|f(0)\rangle + |f(1)\rangle|\} + \tag{4.6}$$

$$\frac{1}{8}|1\rangle\left\{|f(0)\rangle\left(1 + e^{(1\cdot2/8)2\pi i} + e^{(1\cdot4/8)2\pi i} + e^{(1\cdot6/8)2\pi i}\right) + \right.$$

$$\left. |f(1)\rangle\left(e^{(1\cdot1/8)2\pi i} + e^{(1\cdot3/8)2\pi i} + e^{(1\cdot5/8)2\pi i} + e^{(1\cdot7/8)2\pi i}\right)\right\} +$$

$$\frac{1}{8}|2\rangle\left\{|f(0)\rangle\left(1 + e^{(2\cdot2/8)2\pi i} + e^{(2\cdot4/8)2\pi i} + e^{(2\cdot6/8)2\pi i}\right) + \right.$$

$$|f(1)\rangle \left(e^{(2\cdot 1/8)2\pi i} + e^{(2\cdot 3/8)2\pi i} + e^{(2\cdot 5/8)2\pi i} + e^{(2\cdot 7/8)2\pi i} \right) \Bigg\} +$$

$$\frac{1}{8}|3\rangle \left\{ |f(0)\rangle \left(1 + e^{(3\cdot 2/8)2\pi i} + e^{(3\cdot 4/8)2\pi i} + e^{(3\cdot 6/8)2\pi i} \right) + \right.$$

$$|f(1)\rangle \left(e^{(3\cdot 1/8)2\pi i} + e^{(3\cdot 3/8)2\pi i} + e^{(3\cdot 5/8)2\pi i} + e^{(3\cdot 7/8)2\pi i} \right) \Bigg\} +$$

$$\frac{1}{8}|4\rangle \left\{ |f(0)\rangle \left(1 + e^{(4\cdot 2/8)2\pi i} + e^{(4\cdot 4/8)2\pi i} + e^{(4\cdot 6/8)2\pi i} \right) + \right.$$

$$|f(1)\rangle \left(e^{(4\cdot 1/8)2\pi i} + e^{(4\cdot 3/8)2\pi i} + e^{(4\cdot 5/8)2\pi i} + e^{(4\cdot 7/8)2\pi i} \right) \Bigg\} +$$

$$\cdots .$$

Consider, for example, the terms in (4.6) which contain the state $|1\rangle$ in the register X. The complex amplitudes in the first parentheses have the phases,

$$0, \quad \pi/2, \quad \pi, \quad (3/2)\pi. \tag{4.7}$$

Consequently, these amplitudes cancel each other. The complex amplitudes in the second parentheses have the phases,

$$\pi/4, \quad 3\pi/4, \quad 5\pi/4, \quad 7\pi/4, \tag{4.8}$$

which also differ by $\pi/2$. So the corresponding complex amplitudes also cancel each other. Below we present the corresponding phases for four states,

$$|2, f(0)\rangle : \quad 0, \quad \pi, \quad 2\pi, \quad 3\pi, \tag{4.9}$$

$$|2, f(1)\rangle : \quad \pi/2, \quad 3\pi/2, \quad 5\pi/2, \quad 7\pi/2,$$

$$|3, f(0)\rangle : \quad 0, \quad 3\pi/2, \quad 3\pi, \quad 9\pi/2,$$

$$|3, f(1)\rangle : \quad 3\pi/4, \quad 9\pi/4, \quad 15\pi/4, \quad 21\pi/4.$$

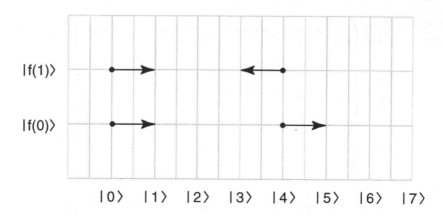

Figure 4.3: The vector-diagram corresponding to the wave function Ψ' (4.12). This diagram is obtained from Fig. 4.2 by addition the vectors (amplitudes) corresponding to the same states, $|i, j\rangle$, when $f(0) = f(2) = f(4) = f(6)$ and $f(1) = f(3) = f(5) = f(7)$. The length of each vector is $1/2$.

For all these functions, the corresponding complex amplitudes in (4.6) cancel each other. For the terms in (4.6) which involve the function $|4, f(0)\rangle$, we have for the corresponding phases,

$$0, \quad 2\pi, \quad 4\pi, \quad 6\pi, \qquad (4.10)$$

which provide a constructive interference for the complex amplitudes. The same is valid for the function $|4, f(1)\rangle$, with the phases,

$$\pi, \quad 3\pi, \quad 5\pi, \quad 7\pi. \qquad (4.11)$$

Finally, from (4.6), we have for Ψ',

$$\Psi' = \frac{1}{2}\{|0, f(0)\rangle + |0, f(1)\rangle + |4, f(0)\rangle + e^{i\pi}|4, f(1)\rangle\}. \quad (4.12)$$

The wave function Ψ' is schematically represented by the vector-diagram, in Fig. 4.3. Now, measuring the state of the register X, we get the numbers 0 or 4. Each of these has the probability $1/2$. According to

Shor's algorithm, a measurement of the state of register X gives one of the values, k,

$$k = 0, \quad D/T, \quad 2D/T, \quad 3D/T, ..., \quad \frac{(T-1)D}{T}, \qquad (4.13)$$

where D is the number of possible digital states of register X (if D is divisible by T). In our case, $D = 2^3 = 8$. Measuring the state of the X register, corresponding to wave function (4.12), one finds that the values of k are: $k = 0$ or $k = 4$. From these measurements, and taking into account (4.13), one concludes that $T = 8/4 = 2$.

The question remains – how to find the period T of the function $f(x)$ if in the process of measurement, the quantum algorithm provides many integer values, k, which are multiples of D/T, where D is the total number of states? Consider a simplified case when D is exactly divisible by T. Assume, for example, that $T = 8$. Let us consider how to get this number, $T = 8$, if the result of measurement of the state of the register X (the values k) is known. Assume that,

$$D = 2^7 = 128. \qquad (4.14)$$

It follows from (4.13), that the measurement of the state of register X gives one of the following 8 values of k,

$$k = 0, \quad 16, \quad 2 \cdot 16 = 32, \quad 3 \cdot 16 = 48, ..., 16 \cdot 7 = 112. \quad (4.15)$$

Let us, for example, suppose that the value k is measured and one obtains the value, 80. In this case we have for D/k,

$$\frac{D}{k} = \frac{128}{80}. \qquad (4.16)$$

We can find the greatest common divisor of 128 and 80, to get,

$$\frac{D}{k} = \frac{8}{5}. \qquad (4.17)$$

The numerator of this fraction is equal to the period T. For other values of k from (4.15) we get,

$$\frac{D}{k} = 8, \quad 4, \quad \frac{8}{3}, \quad 2, \quad \frac{4}{3}, \quad \frac{8}{7}. \qquad (4.18)$$

When we reduce D/k to the lowest terms (see (4.18)), we get the maximum value of the numerator 8, which is equal to the period, T. The probability of getting this value is high enough: W=1/2.

Thus, a quantum measurement of the state of the register X produces,

$$k = m\frac{D}{T}, \quad m = 0, 1, 2, ..., T - 1, \quad (4.19))$$

with equal probability. The fraction,

$$D/k = T/m, \quad (4.20)$$

in the lowest terms has the greatest numerator which is equal to the period T, if T and m do not have common divisors other than one. It is shown that the probability of this event is high enough to provide an efficient calculation with the probability of success as close to one as we wish [39].

Let us briefly summarize the algorithm described above. Using a quantum mechanical approach, we design a wave function, Ψ' (4.5), which involves a superposition of all possible values of the argument x. The quantum computer "tries" all these numbers and automatically selects the superposition of the states with the "desired" measured values of the register X — multiples of D/T. Note, that the quantum algorithm is deterministic, but the output is probabilistic. The main advantage of a quantum computer is that it tries all possible values of x simultaneously (in parallel). But this "quantum parallelism" does not require many computational steps, because the undesirable (incorrect) numbers cancel each other, leaving only the correct values of x. In the case considered above, these correct values appeared with equal probabilities.

Chapter 5

The Discrete Fourier Transform

The quantum algorithm described above includes the use of the discrete Fourier transform (4.4). The question is how to describe this transformation in terms of quantum-mechanical operators? The efficient algorithm for Fourier transform based on application of quantum-mechanical operators was suggested by Coppersmith and Deutsch (see review [39]). Assume we have L qubits in the register X, which can hold any number x, from 0 to $2^L - 1$. Any number x (in decimal notation) can be expressed as the state,

$$|x\rangle = |x_{L-1}x_{L-2}...x_1x_0\rangle = |x_{L-1}\rangle \otimes |x_{L-2}\rangle \otimes ...|x_1\rangle \otimes |x_0\rangle, \quad (5.1)$$

where

$$x = \sum_{i=0}^{L-1} x_i 2^i, \quad (x_i = 0, 1). \quad (5.2)$$

The symbol $\otimes$ in (5.1) means a tensor product, which represents a different notation for L-qubit basic states. In what follows, we shall omit the symbol $\otimes$. Let us introduce the one-qubit (one-atom) operator, A_j, which acts only on the qubit represented by j-th atom. This operator is intended to "mix" in a proper way the two basic states, $|0_j\rangle$ and $|1_j\rangle$ of

31

j-th qubit. The explicit form of the operator A_j is,

$$A_j = 2^{-1/2}(|0_j\rangle\langle 0_j| + |0_j\rangle\langle 1_j| + |1_j\rangle\langle 0_j| - |1_j\rangle\langle 1_j|), \qquad (5.3)$$

$$(j = 0, ..., L - 1).$$

The action of the operator $|i_j\rangle\langle k_j|$ on the state $|n_j\rangle$ is defined by the rule,

$$|i_j\rangle\langle k_j| \cdot |n_j\rangle = \delta_{kn}|i_j\rangle, \qquad (5.4)$$

$$\delta_{kn} = \begin{cases} 1, & k = n, \\ 0, & k \neq n. \end{cases}$$

In matrix representation, the operator $|n_j\rangle\langle m_j|$ only has a non-zero matrix element in the n-th row and m-th column (the rows and columns are counted from zero). For example,

$$|0_j\rangle\langle 1_j| = \begin{vmatrix} 0 & 1 \\ 0 & 0 \end{vmatrix}_j. \qquad (5.5)$$

The index j indicates that the matrix (5.5) acts only on the states of the j-th qubit (state $|x_j\rangle$ in (5.1)).

We shall also introduce a two-qubit operator which acts on the states of qubits, j and k,

$$B_{jk} = |0_{jk}\rangle\langle 0_{jk}| + |1_{jk}\rangle\langle 1_{jk}| + |2_{jk}\rangle\langle 2_{jk}| + e^{i\theta_{jk}}|3_{jk}\rangle\langle 3_{jk}|, \qquad (5.6)$$

$$\theta_{jk} = \frac{\pi}{2^{k-j}}.$$

In (5.6) the following notation is used,

$$|0_{jk}\rangle = |0_j 0_k\rangle, \quad |1_{jk}\rangle = |0_j 1_k\rangle, \qquad (5.7)$$

$$|2_{jk}\rangle = |1_j 0_k\rangle, \quad |3_{jk}\rangle = |1_j 1_k\rangle,$$

Taking into account (5.4), we have the rule of action of the one-qubit operator, A_j, on the basic states of a qubit,

$$A_j|0_j\rangle = \frac{1}{\sqrt{2}}(|0_j\rangle + |1_j\rangle), \qquad (5.8)$$

$$A_j|1_j\rangle = \frac{1}{\sqrt{2}}(|0_j\rangle - |1_j\rangle).$$

Using (5.4), we have from (5.6),

$$B_{jk}|0_{jk}\rangle = |0_{jk}\rangle, \quad B_{jk}|1_{jk}\rangle = |1_{jk}\rangle, \quad B_{jk}|2_{jk}\rangle = |2_{jk}\rangle, \qquad (5.9)$$

$$B_{jk}|3_{jk}\rangle = \exp(i\pi/2^{k-j})|3_{jk}\rangle.$$

It follows from (5.9) that the operator B_{jk} changes only the state

$$|3_{jk}\rangle = |1_j, 1_k\rangle,$$

shifting its phase. The quantum-mechanical operators A_j and B_{jk} allow one to perform a discrete Fourier transform of the wave function. For this, one applies the operator A_{L-1} to the state $|x\rangle$ (5.1). Then, one applies the operator $(A_{L-2}B_{L-2,L-1})$ to the resultant state. After this, one applies the following operators,

$$A_{L-3}B_{L-3,L-2}B_{L-3,L-1}), \qquad (5.10)$$

$$(A_{L-4}B_{L-4,L-3}B_{L-4,L-2}B_{L-4,L-1}), \qquad (5.11)$$

and so on. For our example described by the wave function (4.3), we apply to this wave function the following three groups of operators: A_2, then (A_1B_{12}), then $(A_0B_{01}B_{02})$. As the result, we have:

$$A_0B_{01}B_{02}A_1B_{12}A_2|x\rangle. \qquad (5.12)$$

Assume, for example, that

$$|x\rangle = |2\rangle = |x_2x_1x_0\rangle = |010\rangle, \qquad (5.13)$$

where, on the right side, we omitted the indices which indicate the positions of the qubits because these qubits are written in an ordered form. Then, using (5.8), we have after the first step,

$$A_2|x\rangle = A_2|2\rangle = A_2|x_2\rangle|x_1\rangle|x_0\rangle = \qquad (5.14)$$

$$A_2|0\rangle|1\rangle|0\rangle = \frac{1}{\sqrt{2}}(|0\rangle + |1\rangle)|1\rangle|0\rangle.$$

The subsequent steps give the following results,

$$B_{12}A_2|2\rangle = B_{12} \cdot \frac{1}{\sqrt{2}}(|0\rangle + |1\rangle)|1\rangle|0\rangle = \qquad (5.15)$$

$$\frac{1}{\sqrt{2}}\left\{|0\rangle|1\rangle|0\rangle + e^{i\pi/2}|1\rangle|1\rangle|0\rangle\right\},$$

$$A_1 B_{12} A_2 |2\rangle =$$

$$\frac{1}{2}\left\{|0\rangle(|0\rangle - |1\rangle)|0\rangle + e^{i\pi/2}|1\rangle(|0\rangle - |1\rangle)|0\rangle\right\} =$$

$$\frac{1}{2}\left\{|0\rangle|0\rangle|0\rangle - |0\rangle|1\rangle|0\rangle + \right.$$

$$\left. e^{i\pi/2}|1\rangle|0\rangle|0\rangle - e^{i\pi/2}|1\rangle|1\rangle|0\rangle\right\}.$$

The operator B_{02} does not change the last state in (5.15) because this operator affects only the states $|1k1\rangle$. The operator B_{01} also does not change the last state in (5.15), because it affects only the states $|k11\rangle$.

Finally, from the state $|010\rangle$ (5.13), after applying the operator A_0, we obtain the following state,

$$\frac{1}{\sqrt{8}}\left\{|0\rangle|0\rangle(|0\rangle + |1\rangle) - |0\rangle|1\rangle(|0\rangle + |1\rangle) + \qquad (5.16)\right.$$

$$\left. e^{i\pi/2}|1\rangle|0\rangle(|0\rangle + |1\rangle) - e^{i\pi/2}|1\rangle|1\rangle(|0\rangle + |1\rangle)\right\} =$$

$$\frac{1}{\sqrt{8}}\{(|000\rangle + |001\rangle) - (|010\rangle + |011\rangle) +$$

$$i(|100\rangle + |101\rangle) - i(|110\rangle + |111\rangle)\}.$$

Now we reverse the qubits, to get the final wave function,

$$\frac{1}{\sqrt{8}}\{(|000\rangle + |100\rangle) - (|010\rangle + |110\rangle)+ \qquad (5.17)$$

$$i(|001\rangle + |101\rangle) - i(|011\rangle + |111\rangle)\}.$$

The operation "reverse the qubit" means, for example, for the case of three qubits, that $|ijk\rangle \rightarrow |kji\rangle$. Actually, the quantum-mechanical operation "reverse the qubit" is not applied. Instead of this, one measures the state of register X after the Fourier transform (see Chapter 4), and reads the result of measurement in the opposite order.

In decimal notation, (5.17) is the state,

$$\frac{1}{\sqrt{8}}\{(|0\rangle + |4\rangle) - (|2\rangle + |6\rangle)+ \qquad (5.18)$$

$$i(|1\rangle + |5\rangle) - i(|3\rangle + |7\rangle)\}.$$

It is easy to check that for $|x\rangle = |2\rangle$ the expression (5.18) is equal to (4.4). So, the group of operators (5.12) performs a discrete Fourier transform!

Note that for L qubits the discrete Fourier transform requires L operations A_j, and $[0 + (L - 1)]L/2$ operations, B_{jk}. Thus, the number of computational steps is a quadratic function of L. Consequently, this algorithm is an efficient one.

Chapter 6

Quantum Factorization of Integers

The quantum algorithm for finding the period of a periodic function was used by Shor [19] to factorized an integer. We shall describe this algorithm using as an example, the number, $N = 30$. First, we select randomly a number y, such that the greatest common divisor of the numbers y and N is equal to 1. (It is known that if y ($1 < y < N$) is selected randomly, the probability that two numbers have the greatest common divisor of one unit, is greater than $1/\log_2 N$ [39]. Euclid's efficient algorithm for finding the greatest common divisor is described below.) Now we describe Shor's method of factoring the number N. Let us consider the periodic function,

$$f(x) = y^x \pmod{N}, \quad x = 0, 1, 2, 3, ..., \tag{6.1}$$

where $a \pmod{b}$ is the remainder of a/b. For $N = 30$, let us randomly select $y = 11$. Then, we have from (6.1),

$$f(0) = 1 \pmod{30} = 1, \tag{6.2}$$

$$f(1) = 11 \pmod{30} = 11,$$

$$f(2) = 11^2 \pmod{30} = 1,$$

as $11^2 = 121 = 4 \cdot 30 + 1$. Next,

$$f(3) = 11^3 \,(\mathrm{mod}\,30) = 11, \quad (11^3 = 1331 = 44 \cdot 30 + 11), \quad (6.3)$$

$$f(4) = 11^4 \,(\mathrm{mod}\,30) = 1, \quad (11^4 = 14641 = 488 \cdot 30 + 1).$$

The period T of the function $f(x)$ is, obviously, $T = 2$. This period can be found by the method described in Chapter 4. To find a factor of the number N, we compute $z = y^{T/2} = 11^1 = 11$. The greatest common divisor of $(z + 1, N) = (12, 30)$ is 6. The greatest common divisor of $(z - 1, N) = (10, 30)$ is 10. Both of the numbers 10 and 6 are factors of 30. This is the way to find two factors of a number N if the quantum algorithm provides the period T for the function $f(x)$.

This factoring method fails sometimes. For example, it happens if T is an odd number. It was shown, however, that if y was selected randomly, the probability of failure is small [39]. In particular, in the considered above case, $N = 30$, the function $f(x) = y^x \,(\mathrm{mod}\,30)$ has an even period for any y coprime to 30 $(1 < y < 30)$,

$$T = 2, \quad y = 11, 19, 29, \qquad (6.4)$$

$$T = 4, \quad y = 7, 13, 17, 23.$$

In our calculations, we must find the greatest common divisor of the two numbers, N and y. This can be done efficiently using the Euclid's algorithm. For example, for the numbers 12 and 30, we divide 12 into 30,

$$30 = 2 \cdot 12 + 6. \qquad (6.5)$$

Then, we divide the remainder 6 into the quotient 12: $\frac{12}{6} = 2$. (If the remainder is not equal to zero, we repeat division until the remainder is zero.) The last non-zero remainder (in our example, 6) is the greatest common divisor.

Chapter 7

Logic Gates

Any transformation of bits or qubits can be implemented in hardware using a combination of the simplest logic gates. For a digital computer, the simplest logic gate is the single-bit NOT-gate or N-gate (the gate with one input bit). The truth table for initial (input) a_i and final (output) b_f values is given in Tbl. 7.1. This gate changes the value of a bit,

$$b_f = \bar{a}_i, \quad a_i = 0, 1. \tag{7.1}$$

If we use current circuits, the N-gate can be implemented as shown in Fig. 7.1. In Fig. 7.1, the switch in the upper circuit is closed ($b_f = 1$) when there is no current in the lower circuit ($a_i = 0$). If $a_i = 1$, i.e. the switch in the lower circuit is closed, the magnetic field of the coil opens the switch of the upper circuit, and $b_f = 0$.

The simplest two-bit gates correspond to the Boolean operations of multiplication and addition. (The truth table for Boolean multiplication

a_i	b_f
0	1
1	0

Table 7.1: The truth table for the NOT-gate.

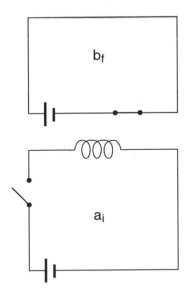

Figure 7.1: The physical implementation of the NOT-gate.

is presented in Tbl. 7.2.) This table corresponds to the truth table of the logical operation "AND" if we consider 0 as "false", and 1 as "true". That is why this logic gate is called the "AND-gate".

The truth table for the second two-bit gate (the Boolean addition) corresponds to the logical operation "OR" (Tbl. 7.3). In figures 7.2 and 7.3 we give an implementation of the gates represented in tables 7.2 and 7.3.

For the AND-gate, the switches of the upper circuit are closed if they arc attracted by the adjacent coils. So, the current in the upper circuit ($c_f = 1$) is possible only if there is a current in both lower circuits ($a_i = b_i = 1$). For the OR-gate, a current in the lower segment ($c_f = 1$) is possible if either of the switches a_i or b_i, or both switches are closed ($a_i = 1$, or $b_i = 1$, or $a_i = b_i = 1$).

With combinations of these three simple gates, we can construct any truth table. For example, assume that we want the EXCLUSIVE-OR (XOR)-gate with the truth table (7.4). This operation is designated by $\oplus$,

a_i	b_i	c_f
0	0	0
0	1	0
1	0	0
1	1	1

$$c_f = a_i b_i$$

Table 7.2: The truth table for the Boolean multiplication (AND-gate).

a_i	b_i	c_f
0	0	0
0	1	1
1	0	1
1	1	1

$$c_f = a_i + b_i$$

Table 7.3: The truth table for Boolean addition (OR-gate).

a_i	b_i	c_f
0	0	0
0	1	1
1	0	1
1	1	0

$$c_f = a_i \oplus b_i$$

Table 7.4: The truth table for the EXCLUSIVE-OR-gate.

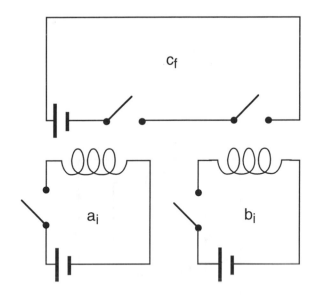

Figure 7.2: The physical implementation of the AND-gate.

$$c_f = a_i \oplus b_i, \qquad (7.2)$$

and corresponds to sum mod 2. It provides a unit value for the output, $c_f = 1$, if only one of the input values is equal to one unit. The output is zero for all other cases.

To design the XOR-gate, we can write the truth table for the XOR-gate in terms of Boolean operations. The two digital rows in Tbl. 7.4 provide $c_f = 1$: the second row with $a_i = 0$, $b_i = 1$, and the third row with $a_i = 1$, $b_i = 0$. The second row corresponds to the product $\bar{a}_i b_i$, and the third one corresponds to $a_i \bar{b}_i$. The final expression is,

$$c_f = \bar{a}_i b_i + a_i \bar{b}_i. \qquad (7.3)$$

For example, if $a_i = 1$ and $b_i = 0$, we have,

$$a_i = 0, \quad \bar{b}_i = 1, \quad \bar{a}_i b_i = 0, \quad a_i \bar{b}_i = 1, \qquad (7.4)$$

$$\bar{a}_i b_i + a_i \bar{b}_i = 1.$$

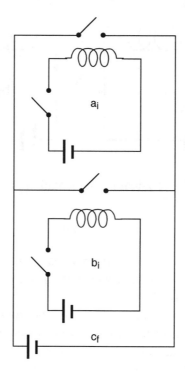

Figure 7.3: The physical implementation of the OR-gate.

Expression (7.3) can be realized using N, OR and AND – gates. If the switch "h" in Fig. 3.2 is closed, and one substitutes A for a_i and B for b_i, the current in the lower circuit will implement the truth table of the XOR-gate.

Chapter 8

Implementation of Logic Gates Using Transistors

We describe here the main ideas of the semiconductor logic gates, following [46]. In conventional computers, transistors are used as the switches. Transistors are made using semiconductors, usually silicon, with a small amount of impurities. If the added impurity introduces a surplus of electrons (n-type semiconductor), we have additional free electrons in the conduction band (an allowed energy region for electrons in which a current can flow). If the added impurity introduces a deficiency of electrons (p-type semiconductor), we get additional "holes" in the valence band which behave like positive charges. If we put n-type and p-type semiconductor slices together, electrons can flow only from the n-type to the p-type semiconductor. So, the system of two slices conducts current only if the negative terminal of the battery is connected to the n-type semiconductor (Fig. 8.1). Silicon is a classical example of semiconductor. It has four valence electrons. Assume, that we add to silicon a small amount of phosphorus which has five valence electrons. In a silicon crystal lattice four of these electrons hold the atom in place in the lattice: each impurity atom bonds four silicon atoms. The fifth electron of phosphorus is free to move. So, we have a surplus of free electrons. This is an n-type semiconductor. If we add boron which

44

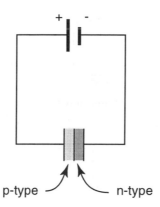

p-type   n-type

Figure 8.1: The system of n-type and p-type semiconductor slices conducts a current only if the negative terminal of the battery is connected to the n-type semiconductor.

has 3 valence electrons it captures an additional electron. This impurity atom bonds to four silicon atoms, but now one has a deficiency of electrons. This is a p-type semiconductor. In a pure silicon, the density of electrons in conduction band, n_e, is equal to the density of holes in the valence band, n_p. Both n_e and n_p depend on the temperature. (The product $n_e n_p$ does not change when we add an impurity.)

Assume that we have a p-type semiconductor and a conducting layer, which is separated from the p-semiconductor by a thin layer of an oxide insulator. In Fig. 8.2, "p" is the p-type silicon, "c" is a conductor, "n" indicates the n-type semiconductors. If we apply a positive voltage to the conducting layer "c", electrons from the n-type semiconductor are attracted to the bottom of the oxide insulator. If we apply a voltage between the n-type semiconductors, we will get a current through the layer formed by these electrons. (These electrons are shown in Fig. 8.2 as a vertical hashed region.) So, we have a switch which can be operated by the voltage $+V$ instead of the magnetic field of the adjacent coil. The n-type semiconductors in this transistor are called the "source" and the "drain," the conductor is called the "gate". This whole system is known

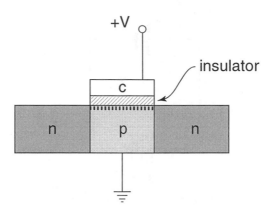

Figure 8.2: Metal-oxide-semiconductor-field-effect-transistor (MOSFET); p is a p-type semiconductor, n are the n-type semiconductors, c is the conductor. Symbol $\overline{\underline{\underline{\ }}}$ means a connection to the ground.

as a MOSFET (metal-oxide-semiconductor-field-effect-transistor). The current between the source and the drain will flow if the voltage between the gate and the source exceeds a critical value. The conventional MOSFET symbol is shown in Fig. 8.3. The potential difference between the drain and the source is positive ($V_d > V_s$). The potential difference between the gate and the source, $V_g - V_s$, as was mentioned above, must exceed some critical value to open the gate. Typically, $V_d - V_s \sim 5V$ and the critical value $V_g - V_s$ is approximately $0.2(V_d - V_s)$. (We describe here only the n-type MOSFET. There also exists a p-type MOSFET.)

The simple scheme of the transistor NOT-gate is shown in Fig. 8.4. Here $+V_{ds}$ is the voltage between the drain and the source, R is the resistor. We suppose that the resistance of the transistor is much less than the resistance of the resistor. If the voltage between the gate and the source, V_{gs}, exceeds the critical value, ($a_i = 1$), current flows through the transistor, and the voltage of the output (i.e. the potential difference between the point "output" and the ground) is approximately equal to zero ($b_f = 0$). In the opposite case ($a_i = 0$), the transistor does not

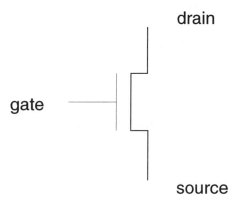

Figure 8.3: The conventional MOSFET symbol.

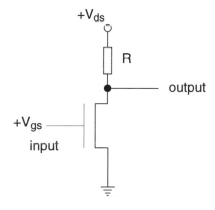

Figure 8.4: The transistor NOT-gate.

a_i	b_i	c_f
0	0	1
0	1	1
1	0	1
1	1	0

$$c_f = \overline{a_i b_i}$$

Table 8.1: The truth table for the NAND-gate.

a_i	b_i	c_f
0	0	1
0	1	0
1	0	0
1	1	0

$$c_f = \overline{a_i + b_i}$$

Table 8.2: The truth table for the NOR-gate

conduct current, and the potential difference between the output and the ground is approximately equal to V_{ds} ($b_f = 1$).

Instead of AND and OR – gates, one can easily design transistor NOT AND (NAND) and NOT OR (NOR) – gates using a couple of transistors. A truth table for the NAND-gate is shown in Tbl. 8.1. Fig. 8.5 shows a transistor implementation of the NAND-gate. As before, $a_i = 1$ means that the potential of the gate (V_{gs}) exceeds the critical value, and the transistor is conducting current. If $a_i = 1$, and $b_i = 1$, both transistors are open, and, consequently, $c_f = 0$ (the potential difference between the points c_f and the ground is very small). For any other case, either both transistors are closed or one of them is closed, $c_f = 1$.

The truth table for the NOR-gate is given in Tbl. 8.2. The transistor realization of the NOR-gate is shown in Fig. 8.6. If $a_i = b_i = 0$ in Fig. 8.6, i.e. the potential at the points a_i and b_i is less than the critical

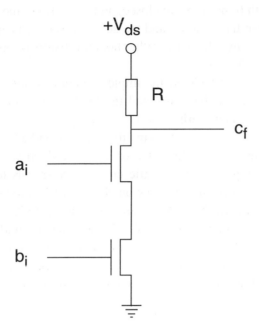

Figure 8.5: The transistor NAND-gate.

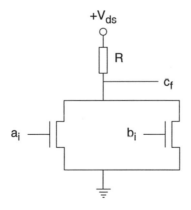

Figure 8.6: The transistor NOR-gate.

value, then both transistors are closed, and $c_f = 1$, i.e. the potential difference between the point c_f and the source (ground) is approximately equal to V_{ds}. In any other case, either both transistors or one of them are open, and $c_f = 0$.

Until now we considered the implementation of logic gates which use different bits for the initial and the final values. For example, in Chapter 7, when considering the N-gate (Fig. 7.1), we have two bits, "a" and "b". The N-gate transforms the value of bit "a" (which we call the initial value, a_i) into the value of bit "b" (which we call the final value, b_f). In Fig. 7.1, one electric circuit corresponds to the bit "a". The other circuit corresponds to the bit "b". In what follows we consider logic gates for which the same circuit corresponds to "a" and "b"; and "b" is the transformed value of "a". These gates are widely used and are important in the theory of quantum computation. For example, the N-gate can operate with only one bit, "a". So, the final value of this bit, a_f, is equal to the complement of the initial value, $a_f = \bar{a}_i$.

Chapter 9

Reversible Logic Gates

A logic gate is called reversible if one can reconstruct the input when one knows the output. For example, the N-gate is reversible. Indeed, if the output $a_f = 0$, we know that the input $a_i = 1$, and vice versa (see Tbl. 7.1, where we should put a_f instead of b_f). The AND-gate is obviously irreversible (see Tbl. 7.2, where we should put a_f instead of c_f). Indeed, if the output $a_f = 0$, we can not say if the pair (a_i, b_i) is equal to (0,0), (0,1), or (1,0). The same is true for OR, XOR, and NOR-gates. Both classical and quantum mechanics in the Hamiltonian formulation describe only reversible processes. So a computer based on quantum-mechanical logic must involve only reversible logic gates. In Tbl. 9.1 we show the truth table for the two-bit CONTROL-NOT (CN) reversible gate. The first bit a is called the control bit. The control bit does not change its value after the action of the CN-gate. The second bit b is called the target bit. The CN-gate changes the value of the target bit if the value of the control bit is equal to one. We can write for the CN-gate,

$$a_f = a_i, \quad b_f = \begin{cases} b_i, & if \quad a_i = 0, \\ \bar{b}_i, & if \quad a_i = 1, \end{cases} \tag{9.1}$$

or

$$b_f = a_i \oplus b_i. \tag{9.2}$$

51

a_i	b_i	a_f	b_f
0	0	0	0
0	1	0	1
1	0	1	1
1	1	1	0

Table 9.1: The truth table for the reversible CN-gate.

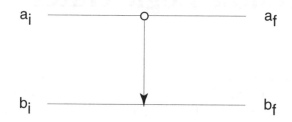

Figure 9.1: The conventional graph for the CN-gate.

It is obvious that information is not lost after the application of the CN-gate: if we know the output a_f and b_f, we can determine the input a_i and b_i. The conventional graph for the CN-gate is shown in Fig. 9.1. The arrow with the circle in Fig. 9.1 shows that the value of b_f depends on the value $a_i = a_f$. In Tbl. 9.2, we show the truth table for the three-bit reversible gate – the CONTROL-CONTROL-NOT (CCN)-gate. The CCN-gate includes two control bits, a and b, which do not change their values, and a target bit c which changes its value only if $a_i = b_i = 1$. The graph for the CCN-gate is shown in Fig. 9.2. The CCN-gate is a universal gate [14]. If we put $a_i = b_i = 1$, then $c_f = \bar{c}_i$, and we have the N-gate. If we put $a_i = 1$, then we get a truth table shown in Tbl. 9.3. One can see that $b_f = b_i$, and $c_f = b_i \oplus c_i$. So we get the CN-gate. If $c_i = 0$, we obtain the truth table represented in Tbl. 9.4. One can see from Tbl. 9.4 that,

$$c_f = a_i b_i, \tag{9.3}$$

a_i	b_i	c_i	a_f	b_f	c_f
0	0	0	0	0	0
0	0	1	0	0	1
0	1	0	0	1	0
0	1	1	0	1	1
1	0	0	1	0	0
1	0	1	1	0	1
1	1	0	1	1	1
1	1	1	1	1	0

$$a_f = a_i, \quad b_f = b_i$$
$$c_f = \begin{cases} \bar{c}_i, & \text{if } a_i = b_i = 1 \\ c_i, & \text{otherwise} \end{cases}$$
or
$$c_f = a_i b_i \oplus c_i$$

Table 9.2: The truth table for the CCN-gate.

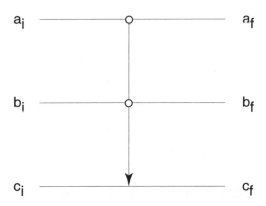

Figure 9.2: The graph for the CCN-gate.

b_i	c_i	a_f	b_f	c_f
0	0	1	0	0
0	1	1	0	1
1	0	1	1	1
1	1	1	1	0

Table 9.3: The truth table for the CCN-gate, if $a_i = 1$.

a_i	b_i	a_f	b_f	c_f
0	0	0	0	0
0	1	0	1	0
1	0	1	0	0
1	1	1	1	1

Table 9.4: The truth table for the CCN-gate, if $c_i = 0$.

	a	b	c	d
CCN(a,b,d)	1	1	1	0
CN(a,b)	1	1	1	1
CCN(b,c,d)	1	0	1	1
CN(b,c)	1	0	1	1
	1	0	1	1

Table 9.5: The consequent values of bits for the operations (9.4).

so we get the AND-gate.

Using a combination of two CCN-gates and two CN-gates, we can create an adder. Indeed, assume that a and b are the bits to be added, and c is the carry-over. Using one additional bit d, we can apply four operations to make an adder,

$$d = 0, \quad \text{CCN}(abd), \quad \text{CN}(ab), \quad \text{CCN}(bcd), \quad \text{CN}(bc). \quad (9.4)$$

At the first step, we set the value of $d = 0$. At the second step, we apply the CCN-gate to the bits a, b, and d (a and b are the control bits, d is the target unit). Then, we apply the CN-gate to bits a and b (a is the control unit, b is the target unit). Then, we apply the CCN-gate to b, c, and d. Finally, we apply the CN-gate to b and c. As a result, the value of the bit c is equal to the sum of bits, and the value of a bit d is the new carry-over.

Let us check, for example, that the sequence of the gates (9.4) provides the adder for the initial values: $a = b = c = 1$. The consequent values of bits are given in Tbl. 9.5. In Tbl. 9.5, the second row shows the initial values of bits. After the action of the CCN(abd)-gate, the value of the target bit d changes because $a = b = 1$. The CN-gate CN(ab) changes the value of b, because $a = 1$. The CCN(bcd)-gate does not change the values of the bits, because one of the control bits $b = 0$. The CN(bc)-gate does not influence the value of bits because the value of the control bit $b = 0$. As a result, we have the correct value of the sum and

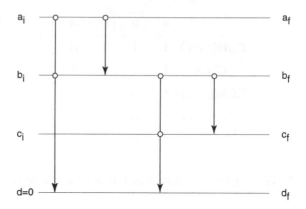

Figure 9.3: Graph for the sequence of operations (9.4).

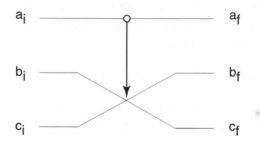

Figure 9.4: The graph for the F-gate.

the carry-over: $c = 1$, $d = 1$. We can sketch the sequence (9.4) using the graph shown in Fig. 9.3. Four arrows in the graph show the action of the CN and CCN – gates.

Finally, in Tbl. 9.6 and Fig. 9.4 we show the truth table and the graph for the well-known three-bit reversible FREDKIN (F)-gate. The F-gate can be called the CONTROL-EXCHANGE-gate. The control bit a_i does not change its value, and the target bits b_i and c_i exchange their values if $a_i = 1$. The F-gate is also a universal gate and can be used to achieve any logical operation [47]. For example, putting $c_i = 0$, we get the truth table shown in Tbl. 9.7. One can see from Tbl. 9.7 that the value of c_f

a_i	b_i	c_i	a_f	b_f	c_f
0	0	0	0	0	0
0	0	1	0	0	1
0	1	0	0	1	0
0	1	1	0	1	1
1	0	0	1	0	0
1	0	1	1	1	0
1	1	0	1	0	1
1	1	1	1	1	1

$$a_f = a_i$$

$$\begin{pmatrix} b_f \\ c_f \end{pmatrix} = \begin{cases} \begin{pmatrix} b_i \\ c_i \end{pmatrix}, & \text{if } a_i = 0 \\ \begin{pmatrix} c_i \\ b_i \end{pmatrix}, & \text{if } a_i = 1 \end{cases}$$

Table 9.6: The truth table for the F-gate.

a_i	b_i	a_f	b_f	c_f
0	0	0	0	0
0	1	0	1	0
1	0	1	0	0
1	1	1	0	1

Table 9.7: The truth table for the F-gate (if $c_i = 0$).

is equal to,

$$c_f = a_i b_i,$$ (9.5)

so we have the AND-gate.

Chapter 10

Quantum Logic Gates

Unlike the digital logic gates, quantum logic gates generally act on a superposition of digital states. Quantum logic gates can be represented by operators or matrices. Consider the matrix A with matrix elements A_{ik}. An adjoint matrix $A^\dagger$ is defined as a matrix with the matrix elements,

$$(A^\dagger)_{ik} = A^*_{ki}, \tag{10.1}$$

where "star" means the complex conjugate. For example, for the matrices,

$$A = \begin{pmatrix} 0 & i \\ i & 0 \end{pmatrix}, \quad B = \begin{pmatrix} 0 & -i \\ i & 0 \end{pmatrix}, \tag{10.2}$$

the adjoint matrices are,

$$A^\dagger = \begin{pmatrix} 0 & -i \\ -i & 0 \end{pmatrix}, \quad B^\dagger = \begin{pmatrix} 0 & -i \\ i & 0 \end{pmatrix}, \tag{10.3}$$

as

$$(A^\dagger)_{12} = A^*_{21} = -i, \quad (A^\dagger)_{21} = A^*_{12} = -i,$$
$$(B^\dagger)_{12} = B^*_{21} = -i, \quad (B^\dagger)_{21} = B^*_{12} = i.$$

Both matrices, A and B, have special properties. For matrix B, we have $B^\dagger = B$. Matrices which are equal to their adjoints are called Hermitian. Hermitian matrices represent physical quantities which can

59

be measured experimentally, i.e. energy, projection of spin (internal angular momentum), projection of magnetic moment, etc. In particular, the matrix $(1/2)B$ describes the y-component of the electron's or the proton's spins.

For both matrices A and B, we have the important equalities,

$$A^\dagger A = AA^\dagger = E, \quad B^\dagger B = BB^\dagger = E \qquad (10.4)$$

where E is the unit matrix,

$$E = \begin{pmatrix} 1 & 0 \\ 0 & 1 \end{pmatrix}. \qquad (10.5)$$

One can check (10.4) using the definition that the product of any two matrices A and B is,

$$(AB)_{ik} = A_{in}B_{nk}, \qquad (10.6)$$

where a summation is assumed over the repeated index n. Thus,

$$(A^\dagger A)_{11} = A^\dagger_{11}A_{11} + A^\dagger_{12}A_{21} = 0 \cdot 0 + (-i) \cdot i = 1, \qquad (10.7)$$

$$(A^\dagger A)_{22} = A^\dagger_{21}A_{12} + A^\dagger_{22}A_{22} = (-i) \cdot i + 0 \cdot 0 = 1,$$

$$(A^\dagger A)_{12} = A^\dagger_{11}A_{12} + A^\dagger_{12}A_{22} = 0,$$

$$(A^\dagger A)_{21} = A^\dagger_{21}A_{11} + A^\dagger_{22}A_{21} = 0.$$

The matrices which satisfy equation (10.4) are called unitary matrices. The time-evolution of quantum-mechanical systems is represented by a unitary matrix. So, quantum logic gates can be represented by unitary matrices (operators).

Consider, for example, the quantum N-gate. It transforms the ground state $|0\rangle$ into the excited state $|1\rangle$, and vice versa, similar to the digital N-gate. For the superpositional state, the N-gate provides the transformation,

$$N \cdot (c_0|0\rangle + c_1|1\rangle) = c_0|1\rangle + c_1|0\rangle. \qquad (10.8)$$

Here c_0 and c_1 are the complex amplitudes of the states. For the initial state $\Psi = (c_0|0\rangle + c_1|1\rangle)$, $|c_0|^2$ is the probability of finding the system in the state $|0\rangle$, and $|c_1|^2$ is the probability of finding the system in the state $|1\rangle$. After action of the N-gate, $|c_0|^2$ is the probability of finding the system in the state $|1\rangle$, and $|c_1|^2$ is the probability of finding the system in the state $|0\rangle$. Of course, $|c_0|^2 + |c_1|^2 = 1$.

If we represent the states $|0\rangle$ and $|1\rangle$ in the form of a column matrix,

$$|0\rangle = \begin{pmatrix} 1 \\ 0 \end{pmatrix} \equiv \alpha, \quad |1\rangle = \begin{pmatrix} 0 \\ 1 \end{pmatrix} \equiv \beta, \tag{10.9}$$

then the N-gate can be represented by the matrix,

$$N = \begin{pmatrix} 0 & 1 \\ 1 & 0 \end{pmatrix}. \tag{10.10}$$

Note, that this matrix is unitary and Hermitian. (The matrix $(1/2)N$ describes the x-component of the electron's or proton's spins.) One can check that,

$$N\alpha = \beta, \quad N\beta - \alpha. \tag{10.11}$$

Indeed, for any square matrix R and column matrix ρ, $R\rho$ is the column matrix with matrix elements,

$$(R\rho)_i = R_{in}\rho_n. \tag{10.12}$$

For example,

$$(N\alpha)_1 = N_{11}\alpha_1 + N_{12}\alpha_2 = 0 + 0 = 0, \tag{10.13}$$

$$(N\alpha)_2 = N_{21}\alpha_1 + N_{22}\alpha_2 = 1 \cdot 1 + 0 = 1.$$

The equations (10.13) are equivalent to the first equation in (10.11), as $\beta_1 = 0$, and $\beta_2 = 1$.

Instead of square matrices like (10.10), we can represent a quantum gate as a sum of the so-called Hubbard operators, X^{ik} ($i, k = 1, 2$). The operator X^{ik} is the square matrix which has the unit matrix element at the intersection of the i-th row and the k-th column. All other matrix elements are equal to zero. For example,

$$X^{11} = \begin{pmatrix} 1 & 0 \\ 0 & 0 \end{pmatrix}, \quad X^{12} = \begin{pmatrix} 0 & 1 \\ 0 & 0 \end{pmatrix}. \tag{10.14}$$

These matrices are convenient for multiplication, because they satisfy the simple rule,

$$X^{ik} X^{mn} = X^{in} \delta_{km}, \tag{10.15}$$

where

$$\delta_{km} = \begin{cases} 1, & k = m, \\ 0, & k \neq m \end{cases}.$$

For example,

$$X^{12} X^{21} = X^{11}, \quad X^{12} X^{11} = 0, \tag{10.16}$$

and so on. In terms of Hubbard operators the N-gate can be represented as,

$$N = X^{12} + X^{21}. \tag{10.17}$$

One can also represent a quantum gate in Dirac notation. In this notation, the matrix X^{ik} has the form,

$$X^{ik} = |i - 1\rangle\langle k - 1|, \tag{10.18}$$

So, the operator X^{11} corresponds to $|0\rangle\langle 0|$. The action of the operator (10.18) can be found using the simple rule,

$$\langle i|k\rangle = \delta_{ik}. \tag{10.19}$$

According to this rule, multiplication of a square matrix $|i\rangle\langle k|$ and a matrix-column $|n\rangle$ is given by the expression,

$$|i\rangle\langle k|n\rangle = |i\rangle \delta_{kn}. \tag{10.20}$$

Multiplication of the square matrices is given by the expression,

$$|i\rangle\langle k|m\rangle\langle n| = |i\rangle\langle n|\delta_{km}. \tag{10.21}$$

In Dirac notation, we have for the N-gate,

$$N = |0\rangle\langle 1| + |1\rangle\langle 0|. \tag{10.22}$$

The first term in (10.22) is responsible for the transformation $|1\rangle \rightarrow |0\rangle$, and the second term is responsible for the inverse transformation: $|0\rangle \rightarrow |1\rangle$,

$$N|1\rangle = |0\rangle\langle 1|1\rangle + |1\rangle\langle 0|1\rangle = |0\rangle, \qquad (10.23)$$

$$N|0\rangle = |0\rangle\langle 1|0\rangle + |1\rangle\langle 0|0\rangle = |1\rangle.$$

It is easy to check that the N operator is unitary. Indeed, one has in Dirac notation,

$$NN\dagger = (|0\rangle\langle 1| + |1\rangle\langle 0|)(|1\rangle\langle 0| + |0\rangle\langle 1|) = \qquad (10.24)$$

$$|0\rangle\langle 0| + |1\rangle\langle 1| = E.$$

In (10.24) we used that the operator $|i\rangle\langle j|$ is adjoint to the operator $|j\rangle\langle i|$.

Chapter 11

Two and Three Qubit Quantum Logic Gates

The quantum CONTROL-NOT (CN) logic gate can be described by the following operator,

$$\text{CN} = |00\rangle\langle00| + |01\rangle\langle01| + |10\rangle\langle11| + |11\rangle\langle10|. \qquad (11.1)$$

The CN-gate is a two-qubit operator where the first qubit is the control and the second qubit is the target. If the control qubit is in the ground state $|0\rangle$, the target qubit does not change its value after the action of the CN-gate. This situation is described by the first two terms in (11.1). In the opposite case, the target qubit changes its value. This case corresponds to the third and forth terms in (11.1). The CN operator, like the N operator, is unitary and Hermitian,

$$(\text{CN})^\dagger = \text{CN}, \quad \text{CN} \cdot (\text{CN})^\dagger = |00\rangle\langle00| + |01\rangle\langle01| + \qquad (11.2)$$

$$|10\rangle\langle10| + |11\rangle\langle11| = E.$$

In decimal notation, the CN-gate can be written as,

$$\text{CN} = |0\rangle\langle0| + |1\rangle\langle1| + |2\rangle\langle3| + |3\rangle\langle2|, \qquad (11.3)$$

64

where

$$|00\rangle \rightarrow |0\rangle, \quad |01\rangle \rightarrow |1\rangle,$$

$$|10\rangle \rightarrow |2\rangle, \quad |11\rangle \rightarrow |3\rangle.$$

The matrix form of the CN-gate in decimal notation is,

$$\text{CN} = \begin{pmatrix} 1 & 0 & 0 & 0 \\ 0 & 1 & 0 & 0 \\ 0 & 0 & 0 & 1 \\ 0 & 0 & 1 & 0 \end{pmatrix}. \tag{11.4}$$

The matrix element $(\text{CN})_{ik}$ corresponds to the term $|i\rangle\langle k|$, where we count i and k from zero in (11.4). Let us write $|0\rangle, |1\rangle, |2\rangle, |3\rangle$ as the column-matrices,

$$\alpha = \begin{pmatrix} 1 \\ 0 \\ 0 \\ 0 \end{pmatrix}, \quad \beta = \begin{pmatrix} 0 \\ 1 \\ 0 \\ 0 \end{pmatrix}, \quad \gamma = \begin{pmatrix} 0 \\ 0 \\ 1 \\ 0 \end{pmatrix}, \quad \delta = \begin{pmatrix} 0 \\ 0 \\ 0 \\ 1 \end{pmatrix}. \tag{11.5}$$

Then, the action of matrix (11.4) on any column-matrix in (11.5) corresponds to the action of the operator (11.3) on the same state. For example,

$$\text{CN}|2\rangle = (|0\rangle\langle 0| + |1\rangle\langle 1| + |2\rangle\langle 3| + |3\rangle\langle 2|)|2\rangle = \tag{11.6}$$

$$|3\rangle\langle 2|2\rangle = |3\rangle.$$

In matrix notation, the i-th element of $(\text{CN}\gamma)$ is given by the expression,

$$(\text{CN}\gamma)_i = (\text{CN})_{ik}\gamma_k = (\text{CN})_{i2}\gamma_2 = \begin{cases} 1, & i = 3, \\ 0, & i = 0, 1, 2. \end{cases} \tag{11.7}$$

So, it follows from (11.7) that,

$$\text{CN}\gamma = \delta, \tag{11.8}$$

which corresponds to (11.6) in the matrix notation.

The three qubit F-gate can be described by the operator,

$$F = |000\rangle\langle000| + |001\rangle\langle001| + |010\rangle\langle010| + \quad (11.9)$$

$$|011\rangle\langle011| + |100\rangle\langle100| + |101\rangle\langle110| +$$

$$|110\rangle\langle101| + |111\rangle\langle111|.$$

The left qubit in (11.9) is the control qubit. If the control qubit is in the ground state $|0\rangle$, the two target qubits do not change their states. This situation is described by first four terms in (11.9). The second four terms in (11.9) describe the opposite case, with the control qubit in the excited state $|1\rangle$ and the target qubits exchanging their states. For example,

$$F|001\rangle = |001\rangle\langle001|001\rangle = |001\rangle, \quad (11.10)$$

and the state of the qubits does not change. At the same time,

$$F|101\rangle = |110\rangle\langle101|101\rangle = |110\rangle, \quad (11.11)$$

and the target qubits exchange their states. In decimal notation, the F gate can be written as,

$$F = |0\rangle\langle0| + |1\rangle\langle1| + |2\rangle\langle2| + |3\rangle\langle3| + |4\rangle\langle4| + |5\rangle\langle6| + \quad (11.12)$$

$$|6\rangle\langle5| + |7\rangle\langle7|.$$

In matrix representation (11.12) has the form,

$$\begin{pmatrix}
1 & 0 & 0 & 0 & 0 & 0 & 0 & 0 \\
0 & 1 & 0 & 0 & 0 & 0 & 0 & 0 \\
0 & 0 & 1 & 0 & 0 & 0 & 0 & 0 \\
0 & 0 & 0 & 1 & 0 & 0 & 0 & 0 \\
0 & 0 & 0 & 0 & 1 & 0 & 0 & 0 \\
0 & 0 & 0 & 0 & 0 & 0 & 1 & 0 \\
0 & 0 & 0 & 0 & 0 & 1 & 0 & 0 \\
0 & 0 & 0 & 0 & 0 & 0 & 0 & 1
\end{pmatrix} \quad (11.13)$$

This F-gate operator is also unitary and Hermitian. Sometimes it is convenient to represent logic gates in terms of creation ($a^\dagger$) and annihilation (a) operators. These operators have the properties,

$$a^\dagger|0\rangle = |1\rangle, \quad a^\dagger|1\rangle = 0, \qquad (11.14)$$

$$a|0\rangle = 0, \quad a|1\rangle = |0\rangle.$$

The creation operator $a^\dagger$ transforms the ground state $|0\rangle$ into the excited state $|1\rangle$. The annihilation operator a transforms the excited state $|1\rangle$ into the ground state $|0\rangle$. In Dirac notation, we have,

$$a^\dagger = |1\rangle\langle 0|, \quad a = |0\rangle\langle 1|. \qquad (11.15)$$

We assume that the operators a and $a^\dagger$ act only on the left qubit; the operators b and $b^\dagger$ act on the central qubit; and the operators c and $c^\dagger$ act on the right qubit. Then, the F-gate can be represented as,

$$F = E + a^\dagger a(b^\dagger c + bc^\dagger - b^\dagger b - c^\dagger c + 2b^\dagger bc^\dagger c). \qquad (11.16)$$

Let us check, as an example, the action of (11.16) on the state $|001\rangle$. We have,

$$F|001\rangle = E|001\rangle - |001\rangle, \qquad (11.17)$$

because $a|0\rangle = 0$, and only the first term E in (11.16) gives a nonzero result. At the same time,

$$F|101\rangle = E|101\rangle + a^\dagger ab^\dagger c|101\rangle - a^\dagger ac^\dagger c|101\rangle = \qquad (11.18)$$

$$|101\rangle + |110\rangle - |101\rangle = |110\rangle.$$

In this case, only three terms in (11.16) produce nonzero results.

Finally, we consider here a three-qubit CCN-gate,

$$CCN = |000\rangle\langle 000| + |001\rangle\langle 001| + |010\rangle\langle 010| + \qquad (11.19)$$

$$|011\rangle\langle 011| + |100\rangle\langle 100| + |101\rangle\langle 101| +$$

$$|110\rangle\langle 111| + |111\rangle\langle 110|.$$

Two left qubits in (11.19) are the control qubits. The CCN-gate changes the state of the right (target) qubit if both control qubits are in the excited state (the last two terms in (11.19)). It is useful to write the CCN-gate in decimal and in matrix representations. In the decimal representation, using Dirac notation, the operator CCN has the form,

$$CCN = |0\rangle\langle 0| + |1\rangle\langle 1| + |2\rangle\langle 2| + |3\rangle\langle 3| + |4\rangle\langle 4| + \tag{11.20}$$

$$|5\rangle\langle 5| + |6\rangle\langle 7| + |7\rangle\langle 6|.$$

In this decimal matrix representation, the operator CCN has the following matrix elements,

$$\begin{cases} (CCN)_{kk} = 1, & if \quad k = 0, 1, 2, 3, 4, 5 \\ (CCN)_{67} = (CCN)_{76} = 1, \\ (CCN)_{in} = 0, & otherwise. \end{cases} \tag{11.21}$$

Chapter 12

One-Qubit Rotation

Here we shall consider how to implement the simplest logic gate, the N-gate, using a two level quantum system. There exists a great number of quantum systems which can be treated approximately as having only two-levels. We shall consider one of them – the proton spin, $I = 1/2$, in a uniform magnetic field $\vec{B}$ which points in the positive z-direction.

The Schrödinger equation for this system can be written as,

$$i\hbar\dot{\Psi} = \mathcal{H}\Psi, \qquad (12.1)$$

where Ψ is the wave function,

$$\Psi = c_0|0\rangle + c_1|1\rangle.$$

The amplitudes c_0 and c_1 satisfy the normalization condition,

$$|c_0|^2 + |c_1|^2 = 1. \qquad (12.2)$$

The Hamiltonian $\mathcal{H}$ of the system is,

$$\mathcal{H} = -\gamma\hbar B I^z = -\hbar\omega_0 I^z, \qquad (12.3)$$

where $\omega_0 = \gamma B$ is the eigenfrequency of the system; γ is the proton gyromagnetic ratio; I^z is the operator which describes the z-component of the spin $1/2$,

$$I^z = \frac{1}{2}(|0\rangle\langle 0| - |1\rangle\langle 1|). \qquad (12.4)$$

In matrix representation, we have for the operator I^z,

$$I^z = \frac{1}{2}\begin{pmatrix} 1 & 0 \\ 0 & -1 \end{pmatrix}. \tag{12.5}$$

The energy of the ground state $|0\rangle$ is equal to $-\hbar\omega_0/2$. The energy of the excited state is $\hbar\omega_0/2$.

At time t, the general solution of the Schrödinger equation (12.1) can be written as,

$$\Psi(t) = c_0(t)|0\rangle + c_1(t)|1\rangle. \tag{12.6}$$

Substituting (12.6) into (12.1), we obtain,

$$i\hbar(\dot{c}_0|0\rangle + \dot{c}_1|1\rangle) = \tag{12.7}$$

$$-\frac{\hbar\omega_0}{2}(|0\rangle\langle 0| - |1\rangle\langle 1|)(c_0|0\rangle + c_1|1\rangle).$$

From (12.7), we can derive two ordinary differential equations for the amplitudes c_0 and c_1,

$$i\dot{c}_0 = -\frac{\omega_0}{2}c_0, \quad i\dot{c}_1 = \frac{\omega_0}{2}c_1. \tag{12.8}$$

The solution of these equations is,

$$c_0(t) = c_0(0)e^{i\omega_0 t/2}, \quad c_1(t) = c_1(0)e^{-i\omega_0 t/2}. \tag{12.9}$$

We now find the quantum-mechanical averages of the x -, y -, and z - components of the proton spin described by (12.9). These values can be measured in experiments in which many proton spins are prepared in the same state at $t = 0$. The operators I^x and I^y are,

$$I^x = \frac{1}{2}(|0\rangle\langle 1| + |1\rangle\langle 0|), \tag{12.10}$$

$$I^y = \frac{i}{2}(-|0\rangle\langle 1| + |1\rangle\langle 0|).$$

The average value $\langle A \rangle$ of any operator A (physical observable) can be found as,

$$\langle A \rangle = \Psi^\dagger A \Psi. \tag{12.11}$$

In our case, we have the wave function,

$$\Psi(t) = c_0(0)e^{i\omega_0 t/2}|0\rangle + c_1(0)e^{-i\omega_0 t/2}|1\rangle. \tag{12.12}$$

First, we calculate the action of the operator I^x on the wave function $\Psi(t)$,

$$I^x \Psi(t)\rangle = \frac{1}{2}(|0\rangle\langle 1| + |1\rangle\langle 0|)\times \tag{12.13}$$

$$\left(c_0(0)e^{i\omega_0 t/2}|0\rangle + c_1(0)e^{-i\omega_0 t/2}|1\rangle \right) =$$

$$\frac{1}{2}c_0(0)e^{i\omega_0 t/2}|1\rangle + \frac{1}{2}c_1(0)e^{-i\omega_0 t/2}|0\rangle.$$

Next we calculate the time-dependent average value $\langle I^x \rangle(t)$,

$$\langle I^x \rangle(t) = \Psi^\dagger(t)I^x\Psi(t) = \tag{12.14}$$

$$\left(c_0^*(0)e^{-i\omega_0 t/2}\langle 0| + c_1^*(0)e^{i\omega_0 t/2}\langle 1| \right) \times$$

$$\left(\frac{1}{2}c_0(0)e^{i\omega_0 t/2}|1\rangle + \frac{1}{2}c_1(0)e^{-i\omega_0 t/2}|0\rangle \right) =$$

$$\frac{1}{2}\left(c_0^*(0)c_1(0)e^{-i\omega_0 t} + c_1^*(0)c_0(0)e^{i\omega_0 t} \right).$$

We can significantly simplify (12.14). Let us write the complex number $c_0(0)c_1^*(0)$ as,

$$c_0(0)c_1^*(0) = ae^{i\varphi}, \tag{12.15}$$

where a is the modulus and φ is the phase of the complex number. Then, the expression (12.14) can be written in the form,

$$\langle I^x \rangle(t) = a\cos(\omega_0 t + \varphi). \tag{12.16}$$

In the same way, one gets,

$$\langle I^y \rangle = \Psi^\dagger(t) I^y \Psi(t) = \tag{12.17}$$

$$\left(c_0^*(0) e^{-i\omega_0 t/2} \langle 0| + c_1^*(0) e^{i\omega_0 t/2} \langle 1| \right) \times$$

$$\left(\frac{i}{2} c_0(0) e^{i\omega_0 t/2} |1\rangle - \frac{i}{2} c_1(0) e^{-i\omega_0 t/2} |0\rangle \right) =$$

$$\frac{i}{2} \left(c_0(0) c_1^*(0) e^{i\omega_0 t} - c_0^*(0) c_1(0) e^{-i\omega_0 t} \right) =$$

$$-a \sin(\omega_0 t + \varphi).$$

Finally, the average value of $\langle I^z \rangle(t)$ is given by the expression,

$$\langle I^z \rangle(t) = \Psi^\dagger(t) I^z \Psi(t) = \tag{12.18}$$

$$\left(c_0^*(0) e^{-i\omega_0 t/2} \langle 0| + c_1^*(0) e^{i\omega_0 t/2} \langle 1| \right) \times$$

$$\frac{1}{2} \left(c_0(0) e^{i\omega_0 t/2} |0\rangle - c_1(0) e^{-i\omega_0 t/2} |1\rangle \right) =$$

$$\frac{1}{2} (|c_0(0)|^2 - |c_1(0)|^2).$$

As one can see from (12.18) the average value $\langle I^z \rangle(t)$ does not depend on t. Note that the length of the average spin does not change in the process of time evolution,

$$\langle I^x \rangle^2 + \langle I^y \rangle^2 + \langle I^z \rangle^2 = \tag{12.19}$$

$$|c_0(0) c_1(0)|^2 + \frac{1}{4} (|c_0(0)|^4 + |c_1(0)|^4 - 2|c_0(0) c_1(0)|^2) =$$

$$\frac{1}{4} (|c_0(0)|^4 + |c_1(0)|^4 + 2|c_0(0) c_1(0)|^2 = \frac{1}{4},$$

$$\langle I^x \rangle^2 + \langle I^y \rangle^2 = a^2.$$

In (12.19) the normalization condition was used: $|c_0(0)|^2 + |c_1(0)|^2 = 1$. The expressions for $\langle I^x \rangle(t)$, $\langle I^y \rangle(t)$, and $\langle I^z \rangle(t)$ describe the precession of average spin vector $\langle \vec{I} \rangle(t)$ around the direction of the magnetic field. The magnitude of the vector $\langle \vec{I} \rangle(t)$ is $1/2$; the z-component of the vector does not change, and the transverse component rotates in the clockwise direction viewed from the top $(+z)$ with the frequency ω_0.

Let us consider what happens if one applies a transverse circularly polarized magnetic field which is resonant with the precession of the vector $\langle \vec{I} \rangle(t)$. (That is, it has a frequency equal to the precession frequency.) This field has the form,

$$B^x = h \cos \omega t, \qquad B^y = -h \sin \omega t. \tag{12.20}$$

In this case, the Hamiltonian of the system,

$$\mathcal{H} = -\gamma \hbar \vec{B} \vec{I}, \tag{12.21}$$

can be written as,

$$\mathcal{H} = -\hbar \omega_0 I^z - \frac{\gamma \hbar}{2}(B^+ I^- + B^- I^+). \tag{12.22}$$

In (12.22), the following notation is introduced,

$$B^+ = B^x + i B^y = h e^{-i\omega t}, \tag{12.23}$$

$$B^- = B^x - i B^y = h e^{i\omega t},$$

$$I^+ = I^x + i I^y = |0\rangle\langle 1|,$$

$$I^- = I^x - i I^y = |1\rangle\langle 0|.$$

Substituting (12.23) into (12.22), one can obtain the following Hamiltonian,

$$\mathcal{H} = -\frac{\hbar}{2} \Big\{ \omega_0(|0\rangle\langle 0| - |1\rangle\langle 1|) + \tag{12.24}$$

$$\Omega\left(e^{i\omega t}|0\rangle\langle 1| + e^{-i\omega t}|1\rangle\langle 0|\right)\Bigg\},$$

where $\Omega = \gamma h$ is the amplitude of the resonant field measured in frequency units. The frequency Ω is called the Rabi frequency. The Rabi frequency describes transitions between the states $|0\rangle$ and $|1\rangle$, under the action of the resonant field. The characteristic time of these transitions $\tau = \pi/\Omega$ is usually much longer than the period of precession, $2\pi/\omega_0$. Substituting the Hamiltonian (12.24) and the wave function (12.6) into the Schrödinger equation (12.1), we derive the equations for c_0 and c_1,

$$i\dot{c}_0 = -\frac{1}{2}\left(\omega_0 c_0 + \Omega e^{i\omega t} c_1\right), \qquad (12.25)$$

$$i\dot{c}_1 = \frac{1}{2}\left(\omega_0 c_1 - \Omega e^{-i\omega t} c_0\right).$$

These equations involve the time-periodic coefficients, $\exp(\pm i\omega t)$. To derive equations with constant coefficients, we use the following substitution,

$$c_0 = c_0' e^{i\omega t/2}, \qquad (12.26)$$

$$c_1 = c_1' e^{-i\omega t/2}.$$

After substituting (12.26) into (12.25), we obtain equations for c_0' and c_1',

$$i\dot{c}_0' = \frac{1}{2}[(\omega - \omega_0)c_0' - \Omega c_1'], \qquad (12.27)$$

$$i\dot{c}_1' = \frac{1}{2}[-(\omega - \omega_0)c_1' - \Omega c_0'].$$

At the resonant condition, $\omega = \omega_0$, we have from (12.27),

$$i\dot{c}_0' = -\frac{1}{2}\Omega c_1', \qquad (12.28)$$

$$i\dot{c}_1' = -\frac{1}{2}\Omega c_0'.$$

The transformation (12.26) is equivalent to the transition to a system of coordinates which rotates with the resonant magnetic field. In this system of coordinates, the circularly polarized magnetic field becomes a constant transverse field. Also, in this system of coordinates, precession around z-axis is absent. So, we effectively "turn off" the permanent magnetic field which is pointed in z-direction. Thus, in this rotating system of coordinates, we have effectively only the transverse constant magnetic field with amplitude $h = \Omega/\gamma$. Next, we shall omit the "prime" in the expressions for c_0', and c_1'. The general solution of (12.28) can be written as,

$$c_0(t) = c_0(0) \cos \frac{\Omega t}{2} + i c_1(0) \sin \frac{\Omega t}{2}, \tag{12.29}$$

$$c_1(t) = i c_0(0) \sin \frac{\Omega t}{2} + c_1(0) \cos \frac{\Omega t}{2}.$$

Let us assume that at $t = 0$, the spin is in the ground state,

$$c_0(0) = 1, \quad c_1(0) = 0. \tag{12.30}$$

Substituting (12.30) into (12.29), we have,

$$c_0(t) = \cos \frac{\Omega t}{2}, \tag{12.31}$$

$$c_1(t) = i \sin \frac{\Omega t}{2}.$$

If we take the duration of the external resonant field, t_1, to be equal to,

$$t_1 = \frac{\pi}{\Omega}, \tag{12.32}$$

then we have from (12.31),

$$c_0(t_1) = 0, \quad c_1(t_1) = i. \tag{12.33a}$$

It follows from (12.33a) that,

$$|c_0(t_1)|^2 = 0, \quad |c_1(t_1)|^2 = 1. \tag{12.33b}$$

Thus, a pulse of a resonant magnetic field with a duration of π/Ω drives the system from the ground state to the excited state. Such a pulse is called a π-pulse. Conversely, if a spin is initially in the excited state,

$$c_0(0) = 0, \quad c_1(0) = 1, \qquad (12.34a)$$

then after the action of a π-pulse we have,

$$c_0(t_1) = i, \quad c_1(t_1) = 0. \qquad (12.34b)$$

So, a π-pulse drives the spin into the ground state. The π-pulse works as a quantum N-operator – it changes the state of the system from $|0\rangle$ to $|1\rangle$ or from $|1\rangle$ to $|0\rangle$. (The common phase factor, $i = \exp(i\pi/2)$ is not significant for the wave function, because this factor does not affect any observable value.) If we apply a pulse with a different duration, we can drive the quantum system into a superpositional state, creating a so-called one-qubit rotation. For example, with $t_1 = \pi/2\Omega$ ($\pi/2$-pulse) and the initial conditions (12.30), we get from (12.31),

$$c_0(t_1) = \cos\frac{\pi}{4}, \quad c_1(t_1) = i\sin\frac{\pi}{4}, \qquad (12.35)$$

$$|c_0(t_1)|^2 = \frac{1}{2}, \quad |c_1(t_1)|^2 = \frac{1}{2}.$$

It follows from (12.35) that a $\pi/2$-pulse drives the system into a superposition with equal weights of the ground and the excited states. Thus, if we measure the state of the system, we get the state $|0\rangle$ or the state $|1\rangle$ with equal probability, 1/2. The same result is obtained when a $\pi/2$-pulse drives the system from a pure excited state (initial conditions (12.34a)).

Finally, we consider change of the average value of a spin components under the action of a resonant field. Repeating previous calculations we have,

$$\langle I^x \rangle = \frac{1}{2}(c_0^* c_1 + c_1^* c_0), \qquad (12.36)$$

$$\langle I^y \rangle = \frac{i}{2}(c_0 c_1^* - c_0^* c_1),$$

$$\langle I^z \rangle = \frac{1}{2}(|c_0|^2 - |c_1|^2).$$

If the system is initially in the ground state and its dynamics is described by (12.31), then the evolution of the average values of the spin components is given by the expressions,

$$\langle I^x \rangle = 0, \qquad\qquad (12.37)$$

$$\langle I^y \rangle(t) = \frac{1}{2} \sin \Omega t,$$

$$\langle I^z \rangle(t) = \frac{1}{2} \cos \Omega t.$$

Eqs (12.37) describe the precession of the average spin around the x-axis, in the rotating system of coordinates. Initially, at $t = 0$, the "average spin" points in the positive z-direction: $\langle I^z \rangle = 1/2$. The z-component of the average spin decreases, and the y component increases. At any moment, $\langle I^y \rangle^2 + \langle I^z \rangle^2 = 1/4$. After the action of a $\pi/2$-pulse ($\Omega t = \pi/2$), we have,

$$\langle I^y \rangle = \frac{1}{2}, \quad \langle I^z \rangle = 0, \qquad\qquad (12.38)$$

i.e. the average spin points in the positive y-direction (A $\pi/2$-pulse drives the average spin in the transversal plane.) After the action of a π-pulse, we have,

$$\langle I^y \rangle = 0, \quad \langle I^z \rangle = -\frac{1}{2}, \qquad\qquad (12.39)$$

i.e. the average spin then points in the negative z-direction.

Chapter 13

A_j – Transformation

Here we discuss how to realize the operator, A_j,

$$A_j = \frac{1}{\sqrt{2}}(|0_j\rangle\langle 0_j| + |0_j\rangle\langle 1_j| + |1_j\rangle\langle 0_j| - |1_j\rangle\langle 1_j|). \tag{13.1}$$

One should recall that the operators A_j and B_{jk},

$$B_{jk} = |0_j 0_k\rangle\langle 0_j 0_k| + |0_j 1_k\rangle\langle 0_j 1_k| + |1_j 0_k\rangle\langle 1_j 0_k| + \tag{13.2}$$

$$e^{i\theta_{jk}}|1_j 1_k\rangle\langle 1_j 1_k|,$$

are necessary to achieve the discrete Fourier transform (see Chapter 5). (The operator A_j acts only on the j-th qubit and the operator B_{jk} acts only on the j-th and k-th qubits.)

Action of the operator A_j on the state $|0_j\rangle$ produces the state,

$$A_j|0_j\rangle = \frac{1}{\sqrt{2}}(|0_j\rangle + |1_j\rangle). \tag{13.3}$$

The same operator transforms the state $|1\rangle$ into,

$$A_j|1_j\rangle = \frac{1}{\sqrt{2}}(|0_j\rangle - |1_j\rangle). \tag{13.4}$$

Now we will find electromagnetic pulses which physically implement the transformations (13.3) and (13.4). We introduce the rotating reference frame and assume that the rotating magnetic field has a phase shift φ, relative to the reference frame,

$$B_x = h\cos(\omega t + \varphi), \quad B_y = -h\sin(\omega t + \varphi). \tag{13.5}$$

Then, we have for B^+ and B^-,

$$B^+ = he^{-(\omega t+\varphi)}, \quad B^- = he^{(\omega t+\varphi)}. \tag{13.6}$$

Correspondingly, the second term in the Hamiltonian (12.24) transforms into,

$$\Omega\left[e^{i(\omega t+\varphi)}|0\rangle\langle 1| + e^{-i(\omega t+\varphi)}|1\rangle\langle 0| \right]. \tag{13.7}$$

For $\varphi \neq 0$, the substitution (12.26) is equivalent to a transition to the rotating reference frame, which is not connected to the applied electromagnetic field, but has the same angular velocity. the direction of the rotating magnetic field makes an angle φ with respect to the x-direction of the rotating frame. Instead of (12.28) we have the following equations,

$$i\dot{c}_0' = -\frac{1}{2}\Omega e^{i\varphi} c_1', \tag{13.8}$$

$$i\dot{c}_1' = -\frac{1}{2}\Omega e^{-i\varphi} c_0'.$$

Dropping the superscript "prime", we get the following solution of (13.8),

$$c_0(t) = c_0(0)\cos\frac{\Omega t}{2} + ic_1(0)e^{i\varphi}\sin\frac{\Omega t}{2}, \tag{13.9}$$

$$c_1(t) = c_1(0)\cos\frac{\Omega t}{2} + ic_0(0)e^{-i\varphi}\sin\frac{\Omega t}{2}.$$

If we choose $\Omega t = \pi/2$ ($\pi/2$-pulse), and the phase $\varphi = \pi/2$, we shall have from (13.9),

$$c_0(t) = \frac{1}{\sqrt{2}}[c_0(0) - c_1(0)], \tag{13.10}$$

$$c_1(t) = \frac{1}{\sqrt{2}}[c_1(0) + c_0(0)].$$

If the system is initially in the ground state ($c_0(0) = 1$, $c_1(0) = 0$), then, after the action of the pulse we have from (13.10),

$$c_0 = \frac{1}{\sqrt{2}}, \quad c_1 = \frac{1}{\sqrt{2}}. \tag{13.11}$$

If the system is initially in the excited state ($c_0(0) = 0$, $c_1(0) = 1$), then, after the action of this pulse we have,

$$c_0 = -\frac{1}{\sqrt{2}}, \quad c_1 = \frac{1}{\sqrt{2}}. \tag{13.12}$$

Thus, the $\pi/2$-pulse with phase $\pi/2$ provides the transformation,

$$|0\rangle \rightarrow \frac{1}{\sqrt{2}}(|0\rangle + |1\rangle), \quad |1\rangle \rightarrow \frac{1}{\sqrt{2}}(|1\rangle - |0\rangle). \tag{13.13}$$

The second transformation differs from the action of the operator A_j by a sign. The question arises: How can one overcome this sign discrepancy? It can be done, for example, if we introduce a third auxiliary level, $|2_j\rangle$ (see Fig. 13.1). The frequency of transition $|0_j\rangle \leftrightarrow |2_j\rangle$, ω_{02}, is assumed to be different from the frequency of transition $|1_j\rangle \leftrightarrow |2_j\rangle$, ω_{12}. Let us first apply a 2π-pulse with a frequency ω_{12}. If the system is initially in the ground state, its state does not change. If the system is initially in the excited state, $|1_j\rangle$, its transformation in general can be described by equation (13.9), where $c_0 \rightarrow c_2$. After the action of a 2π-pulse, we have,

$$c_1 = -c_1, \quad c_2 = 0. \tag{13.14}$$

Hence a 2π-pulse provides the transformation,

$$|1\rangle \rightarrow -|1\rangle. \tag{13.15}$$

Now we apply a $\pi/2$-pulse, with the frequency, ω_{01}, and a phase $\pi/2$. For an initial ground state, substituting $c_0(0) = 1$, $c_1(0) = 0$ into

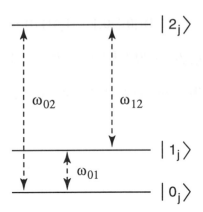

Figure 13.1: The third auxiliary level, $|2_j\rangle$, is used to implement the A_j transformation.

(13.10), we get again (13.11). For an initial excited state, substituting $c_0(0) = 0$, $c_1(0) = -1$ into (13.10), we get,

$$c_0 = \frac{1}{\sqrt{2}}, \quad c_1 = -\frac{1}{\sqrt{2}}. \tag{13.16}$$

Thus, after the action of two pulses we get the desired transformations (13.3), (13.4).

The action of a 2π-pulse, with frequency ω_{12}, on spin j is described by the operator,

$$|0_j\rangle\langle 0_j| - |1_j\rangle\langle 1_j|. \tag{13.17}$$

The action of $\pi/2$-pulses, with frequency ω_{01} and with the $\pi/2$ phase shift, on the same spin, is described by the operator,

$$\frac{1}{\sqrt{2}}(|0_j\rangle\langle 0_j| - |0_j\rangle\langle 1_j| + |1_j\rangle\langle 0_j| + |1_j\rangle\langle 1_j|). \tag{13.18}$$

If we multiply the operator (13.18) by (13.17), we get,

$$\frac{1}{\sqrt{2}}(|0_j\rangle\langle 0_j| - |0_j\rangle\langle 1_j| + |1_j\rangle\langle 0_j| + \tag{13.19}$$

$$|1_j\rangle\langle 1_j|)(|0_j\rangle\langle 0_j| - |1_j\rangle\langle 1_j|) =$$

$$\frac{1}{\sqrt{2}}(|0_j\rangle\langle 0_j| + |0_j\rangle\langle 1_j| + |1_j\rangle\langle 0_j| - |1_j\rangle\langle 1_j|).$$

The operator (13.19) is the operator A_j (see (13.1)).

Chapter 14

B_{jk} – Transformation

Now we discuss how to implement the operator B_{jk} (13.2). Let us, for example, have two interacting three-level systems (see Fig. 14.1). Assume that the energy of the states of the k-atom depends on the state of j-atom: the lower dashed level of the k-atom in Fig. 14.1 corresponds to the state $|1_j\rangle$; the upper "dashed level" corresponds to the state $|0_j\rangle$. So, instead of one frequency $\omega^k(1_k \leftrightarrow 2_k)$, we have two frequencies ω_0^k and ω_1^k (where the subscript corresponds to the state of the neighboring atom, j).

Now, let us apply a π-pulse with frequency ω_1^k to atom k. If atom j is in the ground state, or atom k is in the ground state, or both atoms are in the ground state, a π-pulse does not affect the system. Only if atoms are in the state $|1_j 1_k\rangle$, does the π-pulse drive the k-atom from the state $|1_k\rangle$ to the state $|2_k\rangle$. Let us apply a π-pulse with frequency ω_1^k and phase φ_1, and afterward apply a π-pulse with the same frequency, and phase φ_2. According to (13.9), and substituting $c_0 \to c_2$, we can write the expressions for c_1 and c_2 after the action of a π-pulse,

$$c_1 = ic_2(0)e^{-i\varphi}, \quad c_2 = ic_1(0)e^{i\varphi}, \tag{14.1}$$

where $c_i(0)$ is the value of c_i before the action of the pulse. Assume that the k-atom is initially in the state $|1_k\rangle$ ($c_1 = 1$), and the j-atom is in the state, $|1_j\rangle$. After application of the first π-pulse with frequency ω_1^k and

83

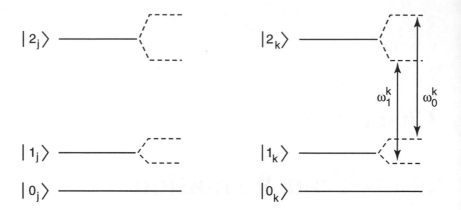

Figure 14.1: Energy levels of two interacting atoms.

phase φ_1, we have

$$c_1 = 0, \quad c_2 = i e^{i\varphi_1}. \tag{14.2}$$

After application of the second π-pulse, with phase φ_2, we get,

$$c_1 = i(i e^{i\varphi_1}) e^{-i\varphi_2} = -e^{i(\varphi_1 - \varphi_2)}, \quad c_2 = 0. \tag{14.3}$$

Thus, the action of two π-pulses is equivalent to (13.2), if

$$\theta_{jk} = \pi + \varphi_1 - \varphi_2.$$

Chapter 15

Unitary Transformations and Quantum Dynamics

We can wonder what the connection is between the quantum dynamics described by the Schrödinger equation and the unitary transformations which describe the quantum logic gates. In this chapter, we shall describe their relation. Let us suppose, for simplicity, that the Hamiltonian of the system is time-independent. Then, the Schrödinger equation,

$$i\hbar\dot{\Psi} = \mathcal{H}\Psi, \tag{15.1}$$

has the solution,

$$\Psi(t) = e^{-i\mathcal{H}t/\hbar}\Psi(0), \tag{15.2}$$

where for any operator F it is assumed,

$$e^{iF} = E + iF + \frac{(iF)^2}{2!} + \frac{(iF)^3}{3!} + \cdots. \tag{15.3}$$

Equation (15.2) defines the unitary transformation of the initial state $\Psi(0)$ into the final state $\Psi(t)$,

$$\Psi(t) = U(t)\Psi(0), \quad U(t) = e^{-i\mathcal{H}t/\hbar}. \tag{15.4}$$

Consider, as an example, a spin 1/2 in a permanent magnetic field, under the action of a resonant electromagnetic pulse. The Hamiltonian of the system is given by Eq. (12.22). We can get the time-independent Hamiltonian using the transformation to the rotating system of coordinates. This transformation can be performed using the formulas,

$$\Psi' = U_r^{\dagger}\Psi, \quad F_t = U_r^{\dagger}FU_r, \tag{15.5}$$

where U_r is the unitary matrix of the transformation in (15.5),

$$U_r = e^{i\omega I^z t}, \tag{15.5a}$$

Ψ' is the wave function in the rotating frame; F is an arbitrary operator in the initial reference frame; F_t is the same operator in the rotating frame; and $\omega = \omega_0$ is the frequency of the rotating magnetic field.

In our case, we make the substitution in (15.1),

$$\Psi = e^{i\omega_0 I^z t}\Psi'.$$

This gives,

$$i\hbar\left(e^{i\omega_0 I^z t}\dot{\Psi}' + i\omega_0 I^z e^{i\omega_0 I^z t}\Psi'\right) = \tag{15.6}$$

$$\left[-\hbar\omega_0 I^z - \frac{\gamma\hbar}{2}\left(B^+ I^- + B^- I^+\right)\right]e^{i\omega_0 I^z t}\Psi'.$$

From (15.6) we get, after simplifications, the Scrödinger equation in the rotating frame,

$$i\hbar\dot{\Psi}' = \mathcal{H}'\Psi', \tag{15.7}$$

$$\mathcal{H}' = -\frac{\gamma\hbar}{2}e^{-i\omega_0 I^z t}(B^+ I^- + B^- I^+)e^{i\omega_0 I^z t}.$$

The right side in Eq. (15.7) describes the interaction of the spin with the electromagnetic field, in the rotating frame.

To simplify the right side of Eq. (15.7), let us find the time-dependent operator,

$$I_t^- = e^{-i\omega_0 I^z t}I^- e^{i\omega_0 I^z t}. \tag{15.8}$$

For this purpose we consider the time derivative,

$$\frac{dI_t^-}{dt} = (-i\omega_0 I^z)e^{-i\omega_0 I^z t}I^- e^{i\omega_0 I^z t} +$$

$$e^{-i\omega_0 I^z t}I^- e^{i\omega_0 I^z t}i\omega_0 I^z. \tag{15.9}$$

Now, using the expressions for the operators I^z (12.4) and I^- (12.23), we obtain,

$$I^z I^- = \frac{1}{2}(|0\rangle\langle 0| - |1\rangle\langle 1|)|1\rangle\langle 0| = \tag{15.10}$$

$$-\frac{1}{2}|1\rangle\langle 0| = -\frac{1}{2}I^-,$$

$$I^- I^z = \frac{1}{2}|1\rangle\langle 0| = \frac{1}{2}I^-.$$

Using (15.10), we can rewrite Eq. (15.9) as follows,

$$\frac{dI_t}{dt} = i\omega_0 I_t^-. \tag{15.11}$$

From (15.11) we have a solution,

$$I_t^- = e^{i\omega_0 t}I^-. \tag{15.12}$$

In the same way, we can show that,

$$I_t^+ = e^{-i\omega_0 I^z t}I^+ e^{i\omega_0 I^z t} = e^{-i\omega_0 t}I^+. \tag{15.13}$$

Substituting (12.23), (15.12) and (15.13) into (15.7) one can see that the Hamiltonian $\mathcal{H}'$ in the rotating frame is time-independent,

$$\mathcal{H}' = -\frac{\hbar}{2}\Omega(|0\rangle\langle 1| + |1\rangle\langle 0|), \tag{15.14}$$

where $\Omega = \gamma h$ is the Rabi frequency.

Now, in the rotating frame, we can use the relations (15.4) for the time-independent Hamiltonian, $\mathcal{H}'$. In this case, the evolution of the system is described by the unitary operator,

$$U(t) = e^{-i\mathcal{H}'t/\hbar}, \tag{15.15}$$

with the time-independent Hamiltonian $\mathcal{H}'$. According to (15.14), the unitary operator $U(t)$ in (15.15) can be written as,

$$U(t) = \exp\left\{\frac{i\Omega}{2}(|0\rangle\langle 1| + |1\rangle\langle 0|)t\right\}. \qquad (15.16)$$

To simplify this expression, let us consider the time derivatives,

$$\frac{dU}{dt} = \frac{i\Omega}{2}(|0\rangle\langle 1| + |1\rangle\langle 0|)U, \qquad (15.17)$$

$$\frac{d^2U}{dt^2} = -\frac{\Omega^2}{4}U.$$

The second equation is valid because of,

$$(|0\rangle\langle 1| + |1\rangle\langle 0|)^2 = (|0\rangle\langle 0| + |1\rangle\langle 1|) = E, \qquad (15.18)$$

where E is the unit matrix. It follows from the second equation in (15.17) that,

$$U(t) = \sum_{i,k=0}^{1}\left(a_{ik}\cos\frac{\Omega t}{2} + b_{ik}\sin\frac{\Omega t}{2}\right)|i\rangle\langle k|, \qquad (15.19)$$

where a_{ik} and b_{ik} are time-independent coefficients. To find these coefficients, we use the initial conditions,

$$U(0) = E = |0\rangle\langle 0| + |1\rangle\langle 1|, \qquad (15.20)$$

$$\left.\frac{dU}{dt}\right|_0 = \frac{i\Omega}{2}(|0\rangle\langle 1| + |1\rangle\langle 0|).$$

The first equation in (15.20) follows from (15.16) and (15.3). The second equation in (15.20) follows from the first equation in (15.17). Substituting (15.19) into (15.20) we get,

$$a_{00} = 1, \quad b_{00} = 0, \qquad (15.21)$$

$$a_{01} = 0, \quad b_{01} = i,$$

$$a_{10} = 0, \quad b_{10} = i,$$

$$a_{11} = 1, \quad b_{11} = 0.$$

The resulting unitary evolution operator is

$$U(t) = \cos\frac{\Omega t}{2}(|0\rangle\langle 0| + |1\rangle\langle 1|) + i \sin\frac{\Omega t}{2}(|0\rangle\langle 1| + |1\rangle\langle 0|), \quad (15.22a)$$

or in matrix representation,

$$U(t) = \begin{pmatrix} \cos \Omega t/2 & i \sin \Omega t/2 \\ i \sin \Omega t/2 & \cos \Omega t/2 \end{pmatrix}. \quad (15.22b)$$

This exactly corresponds to the solution (12.29) of the Schrödinger equation. Using (15.22), we obtain,

$$\Psi(t) = U(t)\Psi(0) = U(t)(c_0(0)|0\rangle + c_1(0)|1\rangle) = \quad (15.23)$$

$$c_0(t)|0\rangle + c_1(t)|1\rangle,$$

where $c_0(t)$ and $c_1(t)$ are given by (12.29).

Chapter 16

Quantum Dynamics at Finite Temperature

So far we have considered an isolated ("pure") quantum system. The same approach is valid for an ensemble of "pure" quantum systems, under the assumption of zero temperature. In reality, this assumption means that the temperature is small in comparison with the energy separation between the considered levels,

$$k_B T \ll \hbar\omega_0,$$

where k_B is the Boltzmann constant, ω_0 is the frequency of transition between the levels of qubits, $|0\rangle$ and $|1\rangle$; and T is the temperature. Gershenfeld, Chuang and Lloyd [28, 29], and Cory, Fahmy and Havel [30] pointed out that the quantum logic gates and quantum computation can be realized also at finite temperature, and even for high temperatures, $k_B T \gg \hbar\omega_0$. This inequality is typical for electron and nuclear spin systems. For example, for a nuclear spin, the typical transition frequency is $\omega_0/2\pi \sim 10^8 \text{Hz}$. So, at room temperature ($T \sim 300K$) one has: $\hbar\omega_0/k_B T \sim 10^{-5}$. That is why we consider in this chapter a high temperature description of quantum systems. Then, using this approach, we will discuss in Chapter 26 the implementation of quantum logic gates at room temperature.

When considering the case of zero temperature, one can assume that the system is prepared initially, for example, in the ground state. To populate this system in the excited state, one usually applies some additional external electromagnetic pulses. As was already mentioned in the Introduction, one can realize quantum logic gates and quantum computation (at least those discussed in the literature) only for a time interval, t, smaller than the characteristic time of relaxation (decoherence), t_R: $t < t_R$. The relaxation processes exist for both a quantum system at zero temperature (due to interactions with the vacuum and other systems) and for the same system (or an ensemble of these systems) at finite temperature. So, for any concrete quantum system, the time t_R is always finite. Then, the question arises: What are the main differences between a quantum system at zero temperature and at finite temperature, when one considers quantum logic gates and quantum computation? Three different situations will now be discussed below.

I. At zero temperature, it is assumed that one can prepare a quantum system in the desired initial state (pure or superpositional). For example, for an individual two-level atom, this initial condition can be the "ground state", $|0\rangle$, the excited state, $|1\rangle$, or any superposition of these two states, $\Psi(0) = c_0(0)|0\rangle + c_1(0)|1\rangle$. The only restriction is, $|c_0(0)|^2 + |c_1(0)|^2 = 1$. Then, during a time interval, t, smaller than the time of relaxation (decoherence), t_R, one can use this system for quantum logic gates and quantum computation. The corresponding dynamics can be described for $t < t_R$ by the Schrödinger equation.

II. One can deal with the same two-level atoms at finite temperature. For example, these atoms can be "colored." They can have energy levels (or some different quantum numbers) that differ from the atoms in the thermal bath. Because of the finite temperature, the "exact" initial conditions are not known for a particular atom. If, for example, the atom is in equilibrium with the atoms of a thermal bath, what is known, is only the probability of finding this atom in the state $|0\rangle$ or $|1\rangle$,

$$P(E_i) = \frac{e^{-E_i/k_B T}}{\sum_{i=0}^{1} e^{-E_i/k_B T}}, \quad (i = 0, 1). \qquad (16.1)$$

In this situation, one cannot implement quantum logic gates or carry out quantum computation, as described in I even if the time of relaxation, t_R is large enough. The wave function approach (the Schrödinger equation), in principal, cannot be applied because one does not know the initial conditions.

III. It was shown in [28]-[30], that one still can realize quantum logic gates and quantum computation using a density matrix approach for an ensemble of atoms, at finite temperature. Speaking very roughly, the main idea is the following. In equilibrium, there always is a difference between the number of atoms populated, for example, in the states $|0\rangle$ and $|1\rangle$. So, if one introduces a new effective density matrix which describes the evolution of the "difference" of atoms in these two states, then it will be equivalent to the density matrix of an effective "pure" quantum system! The situation is more complicated (see Chapter 26), but the idea looks very promising.

The dynamics of an ensemble of atoms at finite temperature can be described by the density matrix introduced by Von Neumann (see, for example, [48]). This approach we shall use in Chapter 26, when describing the dynamics of the quantum logic gates, for time intervals smaller than the time of relaxation (decoherence).

So, we shall discuss in this chapter the evolution not of a single atom at finite temperature, but of an ensemble of atoms. Every atom of this ensemble can still be described by the wave function,

$$\Psi = c_0|0\rangle + c_1|1\rangle. \tag{16.2}$$

First, we introduce the density matrix for an ensemble of atoms which are "prepared" in the same state at zero temperature. Instead of the wave function (16.2), we can consider the density matrix, ρ,

$$\rho = |c_0|^2|0\rangle\langle 0| + c_0 c_1^*|0\rangle\langle 1| + c_1 c_0^*|1\rangle\langle 0| + \tag{16.3a}$$

$$|c_1|^2|1\rangle\langle 1|.$$

In matrix representation, the density matrix (16.3a) has the form,

$$\rho = \begin{pmatrix} \rho_{00} & \rho_{01} \\ \rho_{10} & \rho_{11} \end{pmatrix}, \tag{16.3b}$$

where we define,

$$|c_0|^2 = \rho_{00}, \quad c_0 c_1^* = \rho_{01}, \quad c_1 c_0^* = \rho_{10}, \quad |c_1|^2 = \rho_{11}. \qquad (16.4)$$

The density matrix, ρ, satisfies the operator equation,

$$i\hbar\dot{\rho} = [\mathcal{H}, \rho], \qquad (16.5)$$

where $[\mathcal{H}, \rho]$ is a commutator defined by,

$$[\mathcal{H}, \rho] \equiv \mathcal{H}\rho - \rho\mathcal{H}. \qquad (16.6)$$

For example, for the matrix element ρ_{00} we have the equation,

$$i\hbar\frac{\partial\rho_{00}}{\partial t} = \mathcal{H}_{00}\rho_{00} + \mathcal{H}_{01}\rho_{10} - \rho_{00}\mathcal{H}_{00} - \rho_{01}\mathcal{H}_{10} = \qquad (16.7)$$

$$\mathcal{H}_{01}\rho_{10} - \rho_{01}\mathcal{H}_{10},$$

where we have assumed that the Hamiltonian $\mathcal{H}$ has the form,

$$\mathcal{H} = \sum_{i,k=0}^{1} \mathcal{H}_{ik}|i\rangle\langle k|. \qquad (16.8)$$

Generally, the matrix elements, $\mathcal{H}_{ik}$, depend on time.

Equation (16.7) can be easily derived from the Schrödinger equation. Indeed, the Schrödinger equation can be written in the form,

$$i\hbar\sum_{n=0}^{1}\dot{c}_n|n\rangle = \left(\sum_{i,k=0}^{1}\mathcal{H}_{ik}|i\rangle\langle k|\right)\left(\sum_{p=0}^{1}c_p|p\rangle\right) = \qquad (16.9)$$

$$\sum_{i,k=0}^{1}\mathcal{H}_{ik}c_k|i\rangle.$$

From (16.9) we have the equation for the coefficient c_0,

$$i\hbar\dot{c}_0 = \mathcal{H}_{00}c_0 + \mathcal{H}_{01}c_1. \qquad (16.10)$$

The complex conjugate equation is,

$$-i\hbar \dot{c}_0^* = \mathcal{H}_{00} c_0^* + \mathcal{H}_{10} c_1^*, \tag{16.11}$$

where we took into consideration the fact that the Hamiltonian is a Hermitian operator,

$$\mathcal{H}_{ik} = \mathcal{H}_{ki}^*. \tag{16.12}$$

We now multiply (16.10) by c_0^*, and (16.11) by $-c_0$. Then we add these equations. As a result, we obtain the following equation,

$$i\hbar \frac{\partial}{\partial t}(c_0 c_0^*) = \mathcal{H}_{01} c_1 c_0^* - \mathcal{H}_{10} c_0 c_1^*, \tag{16.13}$$

which coincides with Eq. (16.7).

For an ensemble of atoms at finite temperature, one uses the averaged matrix,

$$\rho = \begin{pmatrix} \langle |c_0|^2 \rangle & \langle c_0 c_1^* \rangle \\ \langle c_1 c_0^* \rangle & \langle |c_1|^2 \rangle \end{pmatrix}, \tag{16.14}$$

which satisfies the same equation (16.5). In the state of the thermodynamic equilibrium, the density matrix is given by the following matrix elements [48],

$$\rho_{kk} = \frac{e^{-E_k/k_B T}}{e^{-E_0/k_B T} + e^{-E_1/k_B T}}, \quad (k = 0, 1), \tag{16.15}$$

$$\rho_{01} = \rho_{10} = 0.$$

In (16.15), E_k is the energy of the k-th level.

From (16.4) and (16.15), one can see the principal difference between the density matrices for an ensemble of atoms which are prepared in the same state at zero temperature and in the state of the thermodynamic equilibrium, at finite temperature. In the case of zero temperature, if both matrix elements, $\rho_{00} \neq 0$ and $\rho_{11} \neq 0$, then ρ_{01} and ρ_{10} are also not equal to zero. At finite temperature one can have, for example: $\rho_{00} \neq 0$, and $\rho_{11} \neq 0$, but $\rho_{01} = \rho_{10} = 0$. The relations,

$$\rho_{00} + \rho_{11} = 1, \quad \rho_{01} = \rho_{10}^*, \tag{16.16}$$

are valid for both zero and finite temperatures. The values ρ_{00} and ρ_{11} for both cases describe the probabilities of occupying the corresponding energy levels.

Now let us consider, as an example, an ensemble of nuclear spins, $I = 1/2$, in a constant magnetic field which points in the positive z-direction. The Hamiltonian of the system is given by (12.3), with two energy levels,

$$E_0 = -\frac{\hbar\omega_0}{2}, \quad E_1 = \frac{\hbar\omega_0}{2}.$$

The density matrix elements in a state of thermal equilibrium, can be found from (16.15),

$$\rho_{00} = \frac{e^{\hbar\omega_0/2k_BT}}{e^{\hbar\omega_0/2k_BT} + e^{-\hbar\omega_0/2k_BT}}, \tag{16.17}$$

$$\rho_{11} = \frac{e^{-\hbar\omega_0/2k_BT}}{e^{\hbar\omega_0/2k_BT} + e^{-\hbar\omega_0/2k_BT}},$$

$$\rho_{01} = \rho_{10} = 0.$$

For the high temperature case, $\hbar\omega_0 \ll k_BT$ (which is especially interesting for quantum computation on electron and nuclear spins), we can expand (16.17) to first order in $\hbar\omega_0/k_BT$,

$$\rho_{00} = \frac{1}{2}(1 + \hbar\omega_0/2k_BT), \quad \rho_{11} = \frac{1}{2}(1 - \hbar\omega_0/2k_BT). \tag{16.18}$$

The expressions (16.18) can be written in operator form,

$$\rho = \frac{1}{2}E + (\hbar\omega_0/2k_BT)I^z, \tag{16.19}$$

where E is the unit matrix and I^z is the operator for the z-component of spin $1/2$ (see (12.4) and (12.5)). The expression (16.19) can also be obtained from the general expression for the density matrix,

$$\rho = \frac{\exp(-\mathcal{H}/k_BT)}{Tr\{\exp(-\mathcal{H}/k_BT)\}}. \tag{16.20}$$

In (16.20), $\mathcal{H} = -\hbar\omega_0 I^z$ is the Hamiltonian of the system (see (12.3)), and Tr means the sum of the diagonal elements of the density matrix.

The first term in (16.19) describes the density matrix at infinite temperature, $T \to \infty$, with equal population of energy levels. The second term in (16.19) describes the first correction due to the finite temperature.

Let us now consider the evolution of the density matrix under the influence of a resonant electromagnetic field with frequency ω_0. Substituting the Hamiltonian (12.24) into the equation for the density matrix (16.5), we derive the equations for the matrix elements,

$$i\hbar\dot{\rho}_{ik} = \mathcal{H}_{in}\rho_{nk} - \rho_{in}\mathcal{H}_{nk}, \qquad (16.21)$$

where

$$\mathcal{H}_{00} = -\frac{\hbar\omega_0}{2}, \quad \mathcal{H}_{11} = \frac{\hbar\omega_0}{2}, \qquad (16.22)$$

$$\mathcal{H}_{01} = -\frac{\hbar\Omega}{2}e^{i\omega_0 t}, \quad \mathcal{H}_{10} = \mathcal{H}_{01}^*,$$

and the summation over the repeated index n is assumed. We now write the explicit equations for the density matrix elements,

$$2i\dot{\rho}_{00} = -\Omega\left(\rho_{10}e^{i\omega_0 t} - \rho_{01}e^{-i\omega_0 t}\right), \qquad (16.23)$$

$$2i\dot{\rho}_{11} = \Omega\left(\rho_{10}e^{i\omega_0 t} - \rho_{01}e^{-i\omega_0 t}\right),$$

$$2i\dot{\rho}_{01} = -2\omega_0\rho_{01} + \Omega e^{i\omega_0 t}(\rho_{00} - \rho_{11}),$$

$$2i\dot{\rho}_{10} = 2\omega_0\rho_{10} + \Omega e^{-i\omega_0 t}(\rho_{00} - \rho_{11}).$$

Note that the second equation in (16.23) can be obtained from the first one, because $\rho_{00} + \rho_{11} = 1$, and consequently,

$$\dot{\rho}_{11} = -\dot{\rho}_{00}. \qquad (16.24)$$

The last equation in (16.23) can be obtained from the third one, because $\rho_{01} = \rho_{10}^*$.

Equations (16.23) include an explicit dependence on time. To derive time-independent equations for the density matrix, we make the substitutions,

$$\rho_{01} = \rho_{01}' e^{i\omega_0 t}, \qquad \rho_{10} = \rho_{10}' e^{-i\omega_0 t}, \qquad (16.25)$$

which is equivalent to a transition to the rotating frame. Omitting a superscript "prime", we derive from (16.23),

$$2i\,\dot{\rho}_{00} = \Omega(\rho_{01} - \rho_{10}), \qquad (16.26)$$

$$2i\,\dot{\rho}_{01} = \Omega(\rho_{00} - \rho_{11}),$$

$$\rho_{10} = \rho_{01}^*, \qquad \rho_{11} = 1 - \rho_{00}.$$

From (16.26), we have a solution,

$$\rho_{00} = a\cos\Omega t + b\sin\Omega t + 1/2, \qquad (16.27)$$

$$\rho_{01} = c + i(b\cos\Omega t - a\sin\Omega t),$$

$$a = \rho_{00}(0) - \frac{1}{2}, \quad b = \frac{\rho_{01}(0) - \rho_{10}(0)}{2i}, \quad c = \frac{\rho_{01}(0) + \rho_{10}(0)}{2}.$$

Note that all coefficients in (16.27) are real. If the initial state of the system is the state of thermal equilibrium, then only the coefficient a differs from zero, and we have in this case,

$$\rho_{00} = \left(\rho_{00}(0) - \frac{1}{2}\right)\cos\Omega t + \frac{1}{2}, \qquad (16.28)$$

$$\rho_{01} = -i\left(\rho_{00}(0) - \frac{1}{2}\right)\sin\Omega t.$$

When $T \to \infty$, we have from (16.17), $\rho_{00}(0) = 1/2$, and the solution (16.28) does not depend on time,

$$\rho_{00} = \frac{1}{2}, \qquad \rho_{01} = 0. \qquad (16.29)$$

So, the time evolution of the system depends only on the initial deviation of the density matrix in (16.19) from $E/2$.

For the initial density matrix (16.19) we have the solution,

$$\rho_{00} = \frac{1}{2}\left(\frac{\hbar\omega_0}{2k_BT}\cos\Omega t + 1\right), \tag{16.30}$$

$$\rho_{01} = -\frac{i\hbar\omega_0}{4k_BT}\sin\Omega t.$$

If we apply a π-pulse, then after the action of the pulse we have,

$$\rho_{00} = \frac{1}{2}\left(1 - \frac{\hbar\omega_0}{2k_BT}\right), \quad \rho_{01} = 0.$$

Note that after the action of the π-pulse, the value of ρ_{00} is equal to the value of $\rho_{11}(0) = 1 - \rho_{00}(0)$.

Roughly, we can think of this state of an ensemble of spins described by the density matrix (16.19) as that of the single spin in the state $|0\rangle$. Similarly, one can think of the state of the ensemble of spins with the density matrix,

$$\rho = \frac{1}{2}E - \frac{\hbar\omega_0}{2k_BT}I^z, \tag{16.31}$$

as that of a single spin in the state $|1\rangle$. The π-pulse drives an ensemble of spins from the state $|i\rangle$ to the state $|k\rangle$, where $i \neq k$, $i, k = 0$ or 1. Note, that unlike pure quantum-mechanical states, we have the transition,

$$|i\rangle \rightarrow |k\rangle, \quad (i \neq k), \tag{16.32}$$

without any phase factor.

The question arises: What corresponds to the superposition of quantum states in an ensemble of spins at finite temperature? To answer this question, let us apply a $\pi/2$-pulse which produces a superposition of quantum states for a "pure" quantum-mechanical system. From (16.30) we have, after the action of $\pi/2$-pulse,

$$\rho_{00} = \frac{1}{2}, \quad \rho_{01} = -\frac{i\hbar\omega_0}{4k_BT}. \tag{16.33}$$

We see from (16.33) that the quantum superposition of pure states corresponds to the appearance of the nondiagonal elements in the density matrix for an ensemble of spins at finite temperature.

Now let us compare the time evolution of averages for a pure quantum-mechanical system and for the ensemble. For the pure state, the evolution of the average spin is given by (12.37). For an ensemble, the average value of any operator A is given by,

$$\langle A \rangle = Tr\{A\rho\}. \tag{16.34}$$

For spin operators I^x, I^y, and I^z ((12.4) and (12.10)), and the density matrix (16.30), we obtain,

$$\langle I^x \rangle = \rho_{ik}I_{ki}^x = \rho_{01}I_{10}^x + \rho_{10}I_{01}^x = \frac{1}{2}(\rho_{01} + \rho_{10}) = 0, \tag{16.35}$$

$$\langle I^y \rangle = \rho_{01}I_{10}^y + \rho_{10}I_{01}^y = \frac{i}{2}(\rho_{01} - \rho_{10}) = \frac{\hbar\omega_0}{4k_BT}\sin\Omega t,$$

$$\langle I^z \rangle = \rho_{00}I_{00}^z + \rho_{11}I_{11}^z = \frac{1}{2}(\rho_{00} - \rho_{11}) = \rho_{00} - \frac{1}{2} = \frac{\hbar\omega_0}{4k_BT}\cos\Omega t.$$

Taking into consideration that, according to (16.35),

$$\langle I^z \rangle(0) = \frac{\hbar\omega_0}{4k_BT}, \tag{16.36}$$

we obtain,

$$\langle I^x \rangle = 0, \tag{16.37}$$

$$\langle I^y \rangle(t) = \langle I^z \rangle(0)\sin\Omega t,$$

$$\langle I^z \rangle(t) = \langle I^z \rangle(0)\cos\Omega t,$$

which is exactly (12.37) , where $\langle I^z \rangle(0) = 1/2$.

To conclude this chapter, we emphasize that there is no exact correspondence between the dynamics of a pure quantum system and an ensemble. One can see from (12.31) that for a pure system,

$$U^\pi|0\rangle = i|1\rangle, \tag{16.38}$$

$$U^{2\pi}|0\rangle = -|0\rangle,$$

$$U^{3\pi}|0\rangle = -i|1\rangle,$$

$$U^{4\pi}|0\rangle = |0\rangle,$$

where $U^{n\pi}$ is the unitary operator that corresponds to the action of $n\pi$-pulse (n is integer). One can see that a π-pulse provides the additional phase shift $i = e^{i\pi/2}$. Also, one can see, that a 2π-pulse does not return a system to the initial state, because of the phase shift, $-1 = e^{i\pi}$. For an ensemble of spins, it follows from (16.30) that after the action of a π-pulse we have,

$$\pi : \quad |0\rangle \rightarrow |1\rangle, \tag{16.39}$$

and after the action of a 2π-pulse we have,

$$2\pi : \quad |0\rangle \rightarrow |0\rangle.$$

In this case, a 2π-pulse returns the ensemble of spins to the initial state.

Chapter 17

Physical Realization of Quantum Computations

Now we consider the physical implementation of quantum computation in a real physical system. The first physical system used for logic gates was the system of cold ions in an ion trap which is very well isolated from the surrounding.

The standard radio frequency (rf) quadrupole trap (the Paul trap) provides a nonstationary quadrupole electric field, in which a charged particle experiences a restoring force for a displacement in any direction of its motion [49]. A single ion can be located at the center of the trap where the rf field is zero. To store several ions, one can use a linear trap with an additional electrostatic potential for axial confinement [50, 51]. A laser beam with a frequency slightly less than the frequency of optical transition in an ion, cools the ions reducing their kinetic energy.

In a linear trap, the spacing between vibrational levels of the ions may exceed the ionic recoil energy from photon emission (the Lamb-Dicke limit). In this limit, the ion system can be cooled to the ground state of its vibrational motion. Then, each ion is localized in a region which is small compared with the wave length of the photon. The distances between adjacent ions are large enough to allow selective laser excitation of any ion.

Cirac and Zoller suggested an implementation of quantum logic gates in this system using the electronic metastable states of the ions, and energy levels of vibrational motion of the center of mass of the ion string [21]. Here we will describe the implementation of quantum computation in ions in an ion trap.

Assume that several ions are placed into the ion trap to form a linear structure. The spacing between the adjacent ions is supposed to be large enough so that the laser beam can drive a single ion inside the trap. Assume that the first excited state of an ion is a metastable one with a long radiative lifetime. By directing a resonant standing wave laser pulse at any particular ion, one provides a single-qubit rotation between the ground state $|0\rangle$ and the metastable electronic state, $|1\rangle$,

$$U^{\alpha}(\varphi)|0\rangle = \cos(\alpha/2)|0\rangle - i e^{i\varphi} \sin(\alpha/2)|1\rangle, \qquad (17.1)$$

$$U^{\alpha}(\varphi)|1\rangle = \cos(\alpha/2)|1\rangle - i e^{-i\varphi} \sin(\alpha/2)|0\rangle,$$

where α is the angle of rotation, and φ is the laser phase. It is assumed that the equilibrium position of the ion coincides with the antinode (the region of maximum amplitude) of the laser standing wave. (Note that the unitary matrix $U^{\alpha}(\varphi)$ is conjugate to the corresponding matrix for a nuclear spin (see (13.9).) For a rectangular laser pulse, $\alpha = \Omega t$, where, as for the spin system, t is the length of a pulse, and Ω is a Rabi frequency (which is proportional to the electric field of a laser beam). Cirac and Zoller also showed how to implement the CN-gate and B_{jk} transformation between any pair of ions, by applying laser pulses. (We shall describe this method in the next chapter.)

Now let us discuss the simplest example: a factorization of the number $N = 4$, using a system of trapped ions. Assume that the X register contains $D = N^2$ states. So, we have $\log_2 16 = 4$ ions for the X register. Assume that the Y register contains N states. So, we have $\log_2 4 = 2$ ions for the Y register. Next, using a digital computer, we select a number y (see Chapter 6), which is coprime to N (the greatest common divisor of y and N is equal to 1). In our case, we have only one such number, $y = 3$. The values of the periodic function (6.1) are:

$$f(x) = 3^x \pmod 4. \qquad (17.2)$$

We have,

$$f(0) = 1 \,(\text{mod}\, 4) = 1, \qquad (17.3)$$

$$f(1) = 3 \,(\text{mod}\, 4) = 3,$$

$$f(2) = 9 \,(\text{mod}\, 4) = 1,$$

$$f(3) = 27 \,(\text{mod}\, 4) = 3,$$

$$f(4) = 81 \,(\text{mod}\, 4) = 1,$$

and so on. Now suppose that we "do not know" the period of the function $f(x)$, and want to find it using Shor's technique (see Chapter 4). The initial state of the system is the ground state,

$$|0000, 00\rangle. \qquad (17.4)$$

The first four ions in the trap refer to the X register. The last two ions refer to the Y register. Next we apply sequentially $\pi/2$-pulses with phase $\pi/2$ to the ions of the X register, to get the state,

$$\Psi = \frac{1}{4}(|0\rangle + |1\rangle)(|0\rangle + |1\rangle)(|0\rangle + |1\rangle)(|0\rangle + |1\rangle)|00\rangle. \qquad (17.5)$$

Next, we apply the CN-gate (11.1) to the last ion of the X register (control qubit), and to the first ion of the Y register (target qubit). Then, we obtain the following state,

$$\Psi_1 = \frac{1}{4}(|0\rangle + |1\rangle)(|0\rangle + |1\rangle)(|0\rangle + |1\rangle)\otimes \qquad (17.6)$$

$$(|0\rangle|0\rangle + |1\rangle|1\rangle)|0\rangle.$$

Finally, applying a π-pulse of the phase $\pi/2$ to the last qubit of the Y register, we have,

$$\Psi_2 = \frac{1}{4}(|0\rangle + |1\rangle)(|0\rangle + |1\rangle)(|0\rangle + |1\rangle)\otimes \qquad (17.7)$$

$$(|0\rangle|0\rangle + |1\rangle|1\rangle)|1\rangle =$$

$$\frac{1}{4}\{|0000, 01\rangle + |0001, 11\rangle + |0010, 01\rangle + |0011, 11\rangle +$$

$$|0100, 01\rangle + |0101, 11\rangle + |0110, 01\rangle + |0111, 11\rangle +$$

$$|1000, 01\rangle + |1001, 11\rangle + |1010, 01\rangle + |1011, 11\rangle +$$

$$|1100, 01\rangle + |1101, 11\rangle + |1110, 01\rangle + |1111, 11\rangle\}.$$

Using decimal notation for X and Y registers, we can rewrite (17.7) in the following form,

$$\Psi_2 = \frac{1}{4}\{|0, 1\rangle + |1, 3\rangle + |2, 1\rangle + |3, 3\rangle + |4, 1\rangle + \qquad (17.8)$$

$$+|5, 3\rangle + |6, 1\rangle + |7, 3\rangle + |8, 1\rangle + |9, 3\rangle + |10, 1\rangle + |11, 3\rangle +$$

$$|12, 1\rangle + |13, 3\rangle + |14, 1\rangle + |15, 3\rangle\}.$$

This is the same superposition, $|x, f(x)\rangle$, as (4.3), for the function (17.2), which should be prepared according to Shor's algorithm, for the discrete Fourier transform. Next, one applies the sequence of operators,

$$A_0 B_{01} B_{02} B_{03} A_1 B_{12} B_{13} A_2 B_{23} A_3, \qquad (17.9)$$

to get the discrete Fourier transform for the X register (see Chapter 5). We recall that the operators A_j and B_{jk} are defined by the following rules,

$$A_j|0_j\rangle = \frac{1}{\sqrt{2}}(|0_j\rangle + |1_j\rangle), \qquad (17.10)$$

$$A_j|1_j\rangle = \frac{1}{\sqrt{2}}(|0_j\rangle - |1_j\rangle),$$

$$B_{jk}|0_k 0_j\rangle = |0_k 0_j\rangle, \quad B_{jk}|0_k 1_j\rangle = |0_k 1_j\rangle,$$

$$B_{jk}|1_k 0_j\rangle = |1_k 0_j\rangle, \quad B_{jk}|1_k 1_j\rangle = \exp(i\pi/2^{k-j})|1_k 1_j\rangle.$$

We count the ions of the X register as, $|x_3 x_2 x_1 x_0\rangle$. (We shall describe later how to realize these operators using electromagnetic pulses.) Now, applying (17.9) to the state Ψ_2, we get for the first term on the right side of (17.7),

$$1. \quad A_3|0000, 01\rangle = \frac{1}{\sqrt{2}}(|0\rangle + |1\rangle)|000\rangle|01\rangle = \qquad (17.11)$$

$$\frac{1}{\sqrt{2}}(|0000, 01\rangle + |1000, 01\rangle) \equiv |S_1\rangle,$$

$$2. \quad B_{23}|S_1\rangle = |S_1\rangle,$$

$$3. \quad A_2|S_1\rangle = \frac{1}{2}\{|0\rangle(|0\rangle + |1\rangle)|00\rangle|01\rangle + |1\rangle(|0\rangle +$$

$$|1\rangle)|00\rangle|01\rangle\} =$$

$$\frac{1}{2}(|0000, 01\rangle + |0100, 01\rangle + |1000, 01\rangle + |1100, 01\rangle) \equiv |S_3\rangle,$$

$$4. \quad B_{13}|S_3\rangle = |S_3\rangle,$$

$$5. \quad B_{12}|S_3\rangle = |S_3\rangle,$$

$$6. \quad A_1|S_3\rangle = \frac{1}{\sqrt{8}}(|0000, 01\rangle +$$

$$|0010, 01\rangle + |0100, 01\rangle + |0110, 01\rangle +$$

$$|1000, 01\rangle + |1010, 01\rangle + |1100, 01\rangle + |1110, 01\rangle) \equiv |S_6\rangle,$$

$$7. \quad B_{03}|S_6\rangle = |S_6\rangle,$$

$$8. \quad B_{02}|S_6\rangle = |S_6\rangle,$$

$$9. \quad B_{01}|S_6\rangle = |S_6\rangle,$$

$$10. \quad A_0|S_6\rangle = \frac{1}{4}(|0000, 01\rangle + |0001, 01\rangle + |0010, 01\rangle +$$

$$|0011, 01\rangle + |0100, 01\rangle + |0101, 01\rangle + |0110, 01\rangle + |0111, 01\rangle +$$

$$|1000, 01\rangle + |1001, 01\rangle + |1010, 01\rangle + |1011, 01\rangle +$$

$$|1100, 01\rangle + |1101, 01\rangle + |1110, 01\rangle + |1111, 01\rangle) \equiv |S_{10}\rangle,$$

where $|S_k\rangle$ denotes the state obtained on the k-th step.

Now we shall repeat the same calculations, for example, for the third term of the right side of (17.7). We obtain,

1. $A_3|0010, 01\rangle = \dfrac{1}{\sqrt{2}}(|0\rangle + |1\rangle)|010\rangle|01\rangle =$ (17.12)

$$\dfrac{1}{\sqrt{2}}(|0010, 01\rangle + |1010, 01\rangle) \equiv |S_1\rangle,$$

2. $B_{23}|S_1\rangle = |S_1\rangle,$

3. $A_2|S_1\rangle = \dfrac{1}{2}\{|0\rangle(|0\rangle +$

$|1\rangle)|10\rangle|01\rangle + |1\rangle(|0\rangle + |1\rangle)|10\rangle|01\rangle\} =$

$$\dfrac{1}{2}(|0010, 01\rangle + |0110, 01\rangle +$$

$|1010, 01\rangle + |1110, 01\rangle) \equiv |S_3\rangle,$

4. $B_{13}|S_3\rangle = \dfrac{1}{2}\Big\{ |0010, 01\rangle + |0110, 01\rangle +$

$e^{i\pi/4}|1010, 01\rangle + e^{i\pi/4}|1110, 01\rangle \Big\} \equiv |S_4\rangle,$

5. $B_{12}|S_4\rangle = \dfrac{1}{2}\Big\{ |0010, 01\rangle + e^{i\pi/2}|0110, 01\rangle +$

$e^{i\pi/4}|1010, 01\rangle + e^{3i\pi/4}|1110, 01\rangle \Big\} \equiv |S_5\rangle,$

6. $A_1|S_5\rangle = \dfrac{1}{\sqrt{8}}\Big\{ |0000, 01\rangle - |0010, 01\rangle +$

$e^{i\pi/2}|0100, 01\rangle - e^{i\pi/2}|0110, 01\rangle +$

$e^{i\pi/4}|1000, 01\rangle - e^{i\pi/4}|1010, 01\rangle + e^{3i\pi/4}|1100, 01\rangle -$

$$\left. e^{3i\pi/4}|1110, 01\rangle \right\} \equiv |S_6\rangle,$$

7. $B_{03}|S_6\rangle = |S_6\rangle,$

8. $B_{02}|S_6\rangle = |S_6\rangle,$

9. $B_{01}|S_6\rangle = |S_6\rangle,$

10. $A_0|S_6\rangle = \dfrac{1}{4}\Big\{ |0000, 01\rangle +$

$$|0001, 01\rangle - |0010, 01\rangle - |0011, 01\rangle +$$

$$e^{i\pi/2}|0100, 01\rangle + e^{i\pi/2}|0101, 01\rangle -$$

$$e^{i\pi/2}|0110, 01\rangle - e^{i\pi/2}|0111, 01\rangle +$$

$$e^{i\pi/4}|1000, 01\rangle + e^{i\pi/4}|1001, 01\rangle -$$

$$e^{i\pi/4}|1010, 01\rangle - e^{i\pi/4}|1011, 01\rangle +$$

$$e^{3i\pi/4}|1100, 01\rangle + e^{3i\pi/4}|1101, 01\rangle -$$

$$\left. e^{3i\pi/4}|1110, 01\rangle - e^{3i\pi/4}|1111, 01\rangle \right\} \equiv |S_{10}\rangle,$$

From expressions (17.11) and (17.12), one can see that constructive interference occurs for the states $|0000, 01\rangle$ and $|0001, 01\rangle$. The constructive interference occurs also for the states $|0000, 11\rangle$ and $|0001, 11\rangle$. Measuring the state of the ions in the X register, one gets the state $|0000\rangle$ or $|0001\rangle$ with equal probability, $1/2$. (We shall describe later how to realize such measurements.) Repeating a few times the whole procedure described in this chapter (applying the proper pulses and measurements of the state of the X register), one gets approximately half of the cases for the first 4 ions to be in the state $|0000\rangle$, and the other half, in the state $|0001\rangle$. Reversing the qubits of the X register (see Chapter 5), one gets the states $|0000\rangle$ and $|1000\rangle$, or, in the decimal notation, $|0\rangle$ and $|8\rangle$. This means that,

$$D/T = 16/T = 8, \tag{17.13}$$

(see Chapter 4), and, consequently, the period T of the function $f(x)$ in (17.2) is, $T = 2$. Now we compute $z = y^{T/2} = 3^1 = 3$. The greatest common divisor of $(z + 1, N) = (4, 4)$ is 1. The greatest common divisor of $(z - 1, N) = (2, 4)$ is 2, which is the factor of 4 which we wanted to find.

Chapter 18

CONTROL-NOT Gate in an Ion Trap

Now we consider how to realize the transformations described in the previous Chapter, by applying the electromagnetic pulses to ions in the ion trap. A qubit consists of the ground state and the long-lived (metastable) excited state of an ion. To realize logic gates, Cirac and Zoller [21] considered two excited degenerate states (states having identical energies) of the n-th ion, $|1_n\rangle$ and $|2_n\rangle$, which could be driven by laser beams of different polarizations, say σ_+ and σ_- (Fig. 18.1). The state $|2_n\rangle$ is used as an auxiliary state.

The evolution of any two-level system is described by the Schrödinger equation. That is why, to explain the dynamics of a specific system, it is often convenient to consider a corresponding "effective" spin system, because the evolution of a spin system can be discussed using the language of precession of the average spin (see Chapter 12). We shall use this approach here.

First consider the CN-gate. Roughly speaking, the main idea of Cirac and Zoller is the following. Assume that the control qubit is spanned by the m-th ion and the target qubit is spanned by the n-th ion. A $\pi/2$-pulse with the frequency of the optical transition ω_0 and a polarization σ_+ acts on the n-th ion. Assume the effective spin $\vec{S}_n$,

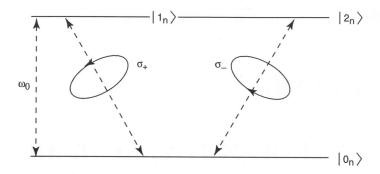

Figure 18.1: Energy levels of the n-th ion. States $|1_n\rangle$ and $|2_n\rangle$ can be driven independently by laser beams with different polarization; ω_0 is the frequency of the optical transition.

associated with this transition, points in the initial state along the $+z$-axis (see Fig. 18.2). After the action of a $\pi/2$-pulse, the spin will be in the $x - y$-plane, and it points, say, along the $+x$-axis, in the rotating frame. Later, one applies the three following pulses with the frequency $\omega_0 - \omega_x$, where ω_x is the vibrational frequency: (1) a π-pulse with σ_+ polarization which acts on the m-th ion, (2) a 2π-pulse with σ_- polarization which acts on the n-th ion, and (3) a π-pulse with σ_+ polarization which acts on the m-th ion. The effect of these three pulses is the following: the direction of the effective spin $\vec{S}_n$ reverses from $+x$ to $-x$-axis, if the m-th ion before the action of these pulses was in the excited state $|1_m\rangle$ (solid line in Fig. 18.2c); the direction of $\vec{S}_n$ does not change if the m-th ion was in the ground state $|0_m\rangle$ (dashed line in Fig. 18.2c). After the action of these three pulses, one applies to the n-th ion a $\pi/2$-pulse with the frequency ω_0, polarization σ_+, and a phase which differs by π from the phase of the first $\pi/2$-pulse. If before the last $\pi/2$-pulse, the spin $\vec{S}_n$ was pointing along the $+x$-axis, it returns to the initial direction along $+z$-axis (Fig. 18.2d – dashed line). If $\vec{S}_n$ was pointing along the $-x$-axis, the spin becomes directed along the $-z$-axis (Fig. 18.2d –

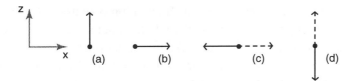

Figure 18.2: Rotation of the effective average spin $\bar{S}_n$, which represent the target qubit: (a) initial direction of the spin; (b) direction of the spin after action of the first $\pi/2$-pulse; (c) direction of the spin after the action of three pulses which implement the Cirac-Zoller gate; (d) direction of the spin after action of the last $\pi/2$-pulse. The solid lines in (c) and (d) correspond to the excited state of the control qubit; the dashed line corresponds to the ground state of the control qubit. In (a) and (b) the direction of the effective spin does not depend on the state of the control qubit.

solid line). So, the n-th ion changes its state if the m-th ion was in the excited state. This is a realization of the quantum CN-gate (the state of m-th ion does not change after the action of five pulses).

To realize this idea, Cirac and Zoller introduced a quantum gate (CZ-gate), which works according the following rules,

$$|0_n 0_m\rangle \rightarrow |0_n 0_m\rangle, \tag{18.1}$$

$$|0_n 1_m\rangle \rightarrow |0_n 1_m\rangle,$$

$$|1_n 0_m\rangle \rightarrow |1_n 0_m\rangle,$$

$$|1_n 1_m\rangle \rightarrow -|1_n 1_m\rangle.$$

Now we consider how to implement this gate. Assume that the laser frequency is $\omega' = \omega_0 - \omega_x$, the polarization is σ_+, and the equilibrium position of the n-th ion coincides with the node of the laser standing wave. Then, the Hamiltonian which describes the interaction between the n-th ion and the laser beam is [21],

$$\mathcal{H}_n = \hbar(\eta\Omega/2)\left(|1_n\rangle\langle 0_n|a e^{-i\varphi} + |0_n\rangle\langle 1_n|a^\dagger e^{i\varphi}\right). \tag{18.2}$$

Here, $a^\dagger$ and a are the creation and annihilation operators of vibrational phonons. The operator $a^\dagger$ drives the whole system of ions from the vibrational ground state to the first excited vibrational state (generates a phonon). The operator a drives the whole system of ions from the excited vibrational state to the vibrational ground state (absorbs a phonon). The parameter η is given by the expression,

$$\eta = k\sqrt{\frac{\pi\hbar}{m_0 N\omega_x}}\cos\Theta, \qquad (18.3)$$

where k is the wave vector of the laser beam, m_0 is the mass of an ion, N is the number of ions, Θ is the angle between the axis of the motion of the center-of-mass of ions and the direction of propagation of the laser beam. The phase of the laser beam, as before, is designated by φ.

If the frequency of the laser beam is $\omega_0 - \omega_x$, then the laser beam can stimulate two processes. If the n-th ion is in the ground state, $|0_n\rangle$, but the whole system of ions is in the excited vibrational states, the whole system can make a transition to the vibrational ground state releasing the energy, $\hbar\omega_x$. At the same time, the n-th ion absorbs this energy , $\hbar\omega_x$, and the energy of the photon, $\hbar(\omega_0 - \omega_x)$, and transfers to the excited state, $|1_n\rangle$. This process is described by the first term in (18.2). If the n-th atom is initially in the excited state, $|1_n\rangle$, but the system of ions is in the vibrational ground state, then the n-th ion can transfer to the ground state generating a photon with the frequency, $(\omega_0 - \omega_x)$, and a phonon with the frequency, ω_x. (Generation of a phonon means a transition of the whole system of ions from the vibrational ground state to the excited vibrational state.) This process is described by the second term in (18.2).

If the laser beam has a σ_- polarization, and the same frequency, $(\omega_0 - \omega_x)$, the interaction between the n-th ion and the laser beam is described by the Hamiltonian,

$$\mathcal{H}_n = \hbar(\eta\Omega/2)\left(|2_n\rangle\langle 0_n|ae^{-i\varphi} + |0_n\rangle\langle 2_n|a^\dagger e^{i\varphi}\right). \qquad (18.4)$$

In this case, the laser beam stimulates a transition between the states, $|0_n\rangle$ and $|2_n\rangle$ with generation or annihilation of a phonon. Under the action of the laser beam, we have a kind of "rotation" between the states $|0_n1\rangle$ and $|1_n0\rangle$, for σ_+ polarization of the laser beam, and between the states $|0_n2\rangle$ and $|2_n0\rangle$, for σ_- polarization of the laser beam. Here $|0\rangle$ and $|1\rangle$ without the indices indicate the ground state and the first excited state of the vibrational motion, respectively. The transformation for rotation under the action of a laser pulse of σ_+ polarization is given by the expression,

$$|0_n1\rangle \rightarrow \cos(\alpha/2)|0_n1\rangle - ie^{i\varphi}\sin(\alpha/2)|1_n0\rangle, \qquad (18.5)$$

$$|1_n0\rangle \rightarrow \cos(\alpha/2)|1_n0\rangle - ie^{-i\varphi}\sin(\alpha/2)|0_n1\rangle.$$

Here α is the angle of rotation, $\alpha = \eta\Omega\tau$, where τ is the duration of the pulse. The transformation of rotation under the action of a σ_- pulse is given by the same expression as (18.5), but with the substitution of 2_n for 1_n. Note that transformation (18.5) is described by the same unitary operator as a one qubit rotation, (17.1), but the formula for the angle α is different for these two cases. We denote the corresponding operator by $U_n^\alpha(\omega, \sigma, \varphi)$, where n indicates the position of the ion, α is the angle of rotation, ω, σ, and φ are the frequency, the polarization, and the phase of the corresponding laser beam. (We will omit the frequency if $\omega = \omega_0$, the polarization, if $\sigma = \sigma_+$, and the phase, if $\varphi = 0$.)

Now we are ready to describe the implementation of the CZ-gate (18.1) using three pulses with frequency, $\omega' = \omega_0 - \omega_x$. Assume first that a π-pulse with polarization σ_+ and phase $\varphi = 0$ acts on the m-th ion. The corresponding transformation is described by the unitary matrix, $U_m^\pi(\omega')$. Secondly, a 2π-pulse with polarization σ_- and a phase $\varphi = 0$ acts on the n-th ion (the unitary transformation $U_n^{2\pi}(\omega', \sigma_-)$. The third pulse is a π-pulse which provides the transformation, $U_m^\pi(\omega')$. Tbl. 18.1 demonstrates change of the states $|k_m p_n 0\rangle$ under the action of three pulses. One can see from Tbl. 18.1 that the first π-pulse drives the control m-th qubit from the excited state to the ground state, while generating a phonon and changing the phase of the corresponding states by $-\pi/2$. The second 2π-pulse, which acts on the target n-th ion, only changes the phase of the state $|0_m 0_n 1\rangle$ by π, leaving all other states unchanged. Note that this pulse does not affect the state $|0_m 1_n 0\rangle$, because

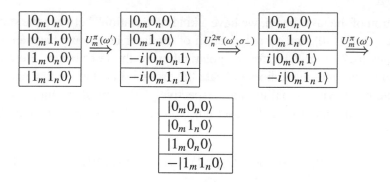

Table 18.1: The CZ-gate, as a result of action of three pulses [21].

of the σ_- polarization of the laser beam. The third π-pulse drives the m-th ion from the ground state to the excited state with an annihilation of a phonon and change of the phase of the corresponding states by $-\pi/2$. As a result, we return to the initial state with the initial phases for all states except for the state $|1_m 1_n 0\rangle$, where we have a phase shift of π. So, the three pulses with the frequency ω' provide the implementation of the CZ-gate.

Now we shall consider the implementation of the CN-gate. As we already mentioned, to provide the CN-gate one applies to the target qubit, n, two additional $\pi/2$ pulses with the resonant frequency ω_0 and the polarization σ_+. The interaction between the resonant field and the n-th ion can be written as,

$$\mathcal{H}_n = (\hbar\Omega/2)\left(|1_n\rangle\langle 0_n|e^{-i\varphi} + |0_n\rangle\langle 1_n|e^{i\varphi}\right). \qquad (18.6)$$

The one qubit rotation, under the action of the laser pulse, is described by the unitary transformation (17.1),

$$|0_n\rangle \rightarrow \cos(\alpha/2)|0_n\rangle - ie^{i\varphi}\sin(\alpha/2)|1_n\rangle, \qquad (18.7)$$

$$|1_n\rangle \rightarrow \cos(\alpha/2)|1_n\rangle - ie^{-i\varphi}\sin(\alpha/2)|0_n\rangle,$$

where $\alpha = \Omega\tau$ is the angle of rotation.

After the first $\pi/2$-pulse with phase $-\pi/2$, one applies three non-resonant pulses to implement the CZ-gate. Then, again one applies a $\pi/2$-pulse with the resonant frequency ω_0, polarization σ_+, but with phase $\pi/2$. The final operator is,

$$U_n^{\pi/2}(\pi/2)U_m^{\pi}(\omega')U_n^{2\pi}(\omega', \sigma_-) \times \qquad (18.8)$$

$$U_m^{\pi}(\omega')U_n^{\pi/2}(-\pi/2).$$

This operator describes the CN-gate. Let us check, for example, the action of (18.8) on the initial state $|1_m 1_n 0\rangle$. After the action of the first pulse (the right-side operator in (18.8)), we have according to (18.7),

$$U_n^{\pi/2}(-\pi/2)|1_m 1_n 0\rangle = |1_m\rangle\left(\frac{1}{\sqrt{2}}|1_n\rangle + \frac{1}{\sqrt{2}}|0_n\rangle\right)|0\rangle = \qquad (18.9)$$

$$\frac{1}{\sqrt{2}}(|1_m 1_n 0\rangle + |1_m 0_n 0\rangle).$$

After the action of three non-resonant pulses we get, according to Tbl. 18.1, the following state,

$$\frac{1}{\sqrt{2}}(-|1_m 1_n 0\rangle + |1_m 0_n 0\rangle). \qquad (18.10)$$

After action of the last resonant $\pi/2$-pulse with the phase $\pi/2$, we obtain, using (18.7),

$$\frac{1}{2}[-|1_m\rangle(|1_n\rangle - |0_n\rangle)|0\rangle + |1_m\rangle(|0_n\rangle + |1_n\rangle)|0\rangle] = \qquad (18.11)$$

$$|1_m 0_n 0\rangle.$$

Thus, under the action of the pulse sequence (18.8), the initial state, $|1_m 1_n 0\rangle$, transforms to the state, $|1_m 0_n 0\rangle$, which corresponds to the action of the CN-gate: If the control qubit, m, is in the state $|1_m\rangle$, then the target qubit changes its state. The whole system of ions remains in the vibrational ground state.

Chapter 19

A_j and B_{jk} Gates in an Ion Trap

In this chapter, we consider how to implement in an ion trap both A_j and B_{jk} gates (17.10), which are necessary for the discrete Fourier transform. First, we discuss the A_j operator. If we apply to the j-th ion a $\pi/2$-pulse with the phase $\pi/2$, we get,

$$|0_j\rangle \rightarrow \frac{1}{\sqrt{2}}(|0_j\rangle + |1_j\rangle), \quad |1_j\rangle \rightarrow \frac{1}{\sqrt{2}}(|1_j\rangle - |0_j\rangle). \quad (19.1)$$

The second transformation differs from the transformation A_j by the sign. To provide the A_j transformation one can first apply a π-pulse with the polarization σ_+ and phase $\pi/2$. Then one applies a 2π-pulse with polarization σ_-. Then, again one applies a π-pulse with polarization σ_+, and phase $-\pi/2$. Finally, one applies a $\pi/2$-pulse with polarization σ_+ and phase $\pi/2$. If the ion j is initially in the ground state, $|0_j\rangle$, then after the action of the first π-pulse, one has the state $|1_j\rangle$. The 2π-pulse does not influence this state because of the σ_- polarization of the laser beam. After the next π-pulse one gets the state $|0_j\rangle$. Finally, after the $\pi/2$-pulse one gets the state $\frac{1}{\sqrt{2}}(|0_j\rangle + |1_j\rangle)$. If the ion j is initially in the excited state $|1_j\rangle$, then one has the following chain of transformations,

$$|1_j\rangle \stackrel{U_j^\pi(\pi/2)}{\Rightarrow} -|0_j\rangle \stackrel{U_j^{2\pi}(\sigma_-)}{\Rightarrow} |0_j\rangle$$

$$\stackrel{U_j^\pi(-\pi/2)}{\Rightarrow} -|1_j\rangle \stackrel{U_j^{\pi/2}(\pi/2)}{\Rightarrow} \frac{1}{\sqrt{2}}(|0_j\rangle - |1_j\rangle). \tag{19.2}$$

Thus, the sequence of 4 pulses provides the implementation of the A_j gate. We described here a scheme for the A_j transformation based on the system with energy levels shown in Fig. 18.1. If one can use an additional energy level, $|3_j\rangle$ and one induces a transition between the levels $|1_j\rangle$ and $|3_j\rangle$ with frequency ω_{13}, then it is more convenient to use the sequence of pulses described in Chapter 13: a 2π-pulse with the frequency ω_{13}, and a resonant $\pi/2$-pulse with the frequency ω_0 and the phase $\pi/2$.

Now let us consider the implementation of the B_{jk} gate in an ion trap. For this, we can use the slightly modified CZ-gate. Instead of a 2π-pulse with σ_- polarization, we take two π-pulses with σ_- polarization and different phases, to provide a phase shift $\pi/2^{k-j}$ for the state $|1_j 1_k 0\rangle$ under the action of the modified CZ-gate. Thus, we apply four pulses with the frequency $\omega' = \omega_0 - \omega_x$: (1) a π-pulse with the polarization σ_+ to the k-th ion, (2) a π-pulse with the polarization σ_- to the j-th ion, (3) a π-pulse with the polarization σ_- and phase φ to the j-th ion, (4) a π-pulse with the polarization σ_+ and phase φ' to the k-th ion. As a result, using (18.5), we get the following transformation,

$$1. \quad U_k^\pi(\omega')|1_k 1_j 0\rangle = -i|0_k 1_j 1\rangle \equiv |S_1\rangle, \tag{19.3}$$

$$2. \quad U_j^\pi(\omega', \sigma_-,)|S_1\rangle = |S_1\rangle,$$

$$3. \quad U_j^\pi(\omega', \sigma_-, \varphi)|S_1\rangle = |S_1\rangle,$$

$$4. \quad U_k^\pi(\omega', \varphi')|S_1\rangle = -e^{i\varphi'}|1_k 1_j 0\rangle.$$

If we put $\varphi' = \pi + \pi/2^{k-j}$, we get the modified CZ transformation,

$$|1_j 1_k 0\rangle \Rightarrow e^{i\pi/2^{k-j}}|1_j 1_k 0\rangle, \tag{19.4}$$

which corresponds to the action of the B_{jk} operator.

Now let us find the transformation for the state $|1_k 0_j 0\rangle$. We have,

$$1. \quad U_k^\pi(\omega')|1_k 0_j 0\rangle = -i|0_k 0_j 1\rangle \equiv |S_1\rangle, \tag{19.5}$$

$$2. \quad U_j^\pi(\omega', \sigma_-)|S_1\rangle = -|0_k 1_j 0\rangle \equiv |S_2\rangle,$$

$$3. \quad U_j^\pi(\omega', \sigma_-, \varphi)|S_2\rangle = ie^{-i\varphi}|0_k 0_j 1\rangle \equiv |S_3\rangle,$$

$$4. \quad U_k^\pi(\omega', \varphi')|S_3\rangle = e^{i(\varphi'-\varphi)}|1_k 0_j 0\rangle \equiv |S_4\rangle.$$

If we put $\varphi = \varphi'$, the state $|1_k 0_j 0\rangle$ does not change under the action of our sequence of pulses. Also, this sequence of pulses does not affect the states $|0_k 0_j 0\rangle$ and $|0_k 1_j 0\rangle)$. Thus, the slightly modified CZ-gate,

$$U_k^\pi(\omega', \varphi)U_j^\pi(\omega', \sigma_-, \varphi)U_j^\pi(\omega', \sigma_-) \times \tag{19.6}$$

$$U_k^\pi(\omega'),$$

with the phase,

$$\varphi = \pi + \frac{\pi}{2^{k-j}},$$

provides the B_{jk} logic gate, which is needed for the discrete Fourier transform.

To conclude this chapter, we discuss the experimental opportunities for realization of quantum computation using ion traps. There exists a number of ions with long-living metastable state of the order of one second, which could be used as qubits for quantum computation. As an example, the Hg^+ ion has a $^2S_{1/2}$ ground state, $|0\rangle$, and a metastable $^2D_{5/2}$ excited state, $|1\rangle$, with the lifetime $\sim 0.1s$. The wave length of the resonant transition is $\lambda_{01} \approx 280\ nm$. The vibrational frequency of the center of mass of the ions in a linear trap is of the order of 1MHz. The system of ions can be cooled by a laser beam with a frequency slightly less than the frequency of the allowed transition. For the Hg^+ ion one can use for this purpose the transition between the ground state and the second excited state, $^2P_{1/2}$ with the wave length, $\lambda_{02} \approx 190\ nm$. The resonant laser beam with the wave length λ_{01} can provide a one-qubit rotation with a Rabi frequency Ω, which depends on the intensity of the laser beam (typically, $\Omega \sim 100kHz$). The combination of the resonant and non-resonant laser beams can provide the CZ and B_{jk} transformations. To measure the state of the ions of the X register, one can use the quantum jump technique [22]. For the Hg^+ ion, for example, a laser beam with the wave length λ_{02} can be applied to the ions of the X register. If the ion fluoresces, the measured state is $|0\rangle$, otherwise, the measured state is $|1\rangle$.

Chapter 20

Linear Chains of Nuclear Spins

A second promising system we consider is the system of nuclear spins which are also well isolated from the surroundings. Let us consider, for example, a solid with the linear chains of atoms (ions) containing nuclear spins. We assume that any interaction between chains is negligible. At the same time, we shall take into consideration the interaction between the nearest neighbors in the chain. Assume that a solid is placed into a uniform magnetic field which is oriented along the z-axis. Then, the one-spin Hamiltonian, without the interaction, can be written in the form (12.3). Following Lloyd [35], we suppose that we have a chain of three types of nuclei, $ABCABCABC$.... All three types have the same spin $I = 1/2$, but they have different magnetic moments (different gyromagnetic ratios). Assume that the interaction between the spins of a chain (for example, a dipole-dipole interaction) is small in comparison with the interaction of spins with the external magnetic field. Then we can take into consideration only the zz part of interaction, $2\hbar J_{k,k+1} I_k^z I_{k+1}^z$, (Ising interaction), which commutes with the non-interacting Hamiltonian. Here, J is the effective constant of the Ising interaction. The Hamiltonian of the whole system (without the electromagnetic field) can be written as,

$$\mathcal{H} = -\hbar \sum_k (\omega_k I_k^z + 2J_{k,k+1} I_k^z I_{k+1}^z), \qquad (20.1)$$

where we sum over all spins in the chain, and

$$\omega_1 = \omega_4 = ... = \omega^A, \qquad (20.2)$$

$$\omega_2 = \omega_5 = ... = \omega^B,$$

$$\omega_3 = \omega_6 = ... = \omega^C.$$

The constant of interaction in (20.1) is also a periodic function of the position k,

$$J_{12} = J_{45} = ... = J^{AB}, \qquad (20.3)$$

$$J_{23} = J_{56} = ... = J^{BC},$$

$$J_{34} = J_{67} = ... = J^{CA}.$$

The Hamiltonian (20.1) does not have off-diagonal terms. The eigenstates of the Hamiltonian represent the spin states of the type,

$$|00111011...\rangle.$$

So, some spins in any eigenstate point "up" (the state $|0\rangle$), and the rest point "down" (the state $|1\rangle$).

Assume that in some state of the system a spin, for example, B, points "up", and in another state of the system this spin , B, points "down", and the directions of all other spins are unchanged. Then the difference, ΔE, between the energies of these two states can have the following values,

$$\Delta E = \hbar(\omega^B \pm J^{AB} \pm J^{BC}). \qquad (20.4)$$

In (20.4), the upper (+) sign for J^{AB} corresponds to the state $|0\rangle$ of the neighboring spin A. The lower (-) sign at J^{AB} corresponds to the state $|1\rangle$ of the neighboring spin, A. The same is true for the sign for J^{BC} and the neighboring spin C. So, we find the following four eigenfrequencies of the Hamiltonian (20.1),

$$\omega_{00}^B = \omega^B + J^{AB} + J^{AC}, \tag{20.5}$$

$$\omega_{01}^B = \omega^B + J^{AB} - J^{AC},$$

$$\omega_{10}^B = \omega^B - J^{AB} + J^{AC},$$

$$\omega_{11}^B = \omega^B - J^{AB} - J^{AC},$$

which correspond to the inversion of one spin, B. In (20.5), ω_{ik}^B means that the left neighbor (A) is in the state $|i\rangle$, and the right neighbor (C) is in the state $|k\rangle$, ($i, k = 0$ or 1).

Let us consider, as an example, how to get the first frequency, ω_{00}^B, in (20.5). Because of the nearest-neighbor interactions, it is enough to consider only three spins and three terms in the Hamiltonian (20.1), and to take into account the inversion of one spin. In our case of inversion of the B spin, we consider a transition for the triplet, ABC,

$$|0_A 0_B 0_C\rangle \leftrightarrow |0_A 1_B 0_C\rangle, \tag{20.6}$$

where the first state refers to the spin A, the second state refers to the spin B, and the third state refers to the spin C. To describe this transition, the only important terms in the Hamiltonian (20.1) are the following,

$$\mathcal{H}' = -\hbar(\omega^B I_B^z + 2J^{AB} I_A^z I_B^z + 2J^{BC} I_B^z I_C^z), \tag{20.7}$$

where the operators I_A^z, I_B^z, and I_C^z act on the corresponding states in (20.6). Using the expression for the operator I^z (12.4), we obtain,

$$\mathcal{H}'|0_A 0_B 0_C\rangle = -\frac{\hbar}{2}(\omega^B + J^{AB} + J^{BC})|0_A 0_B 0_C\rangle, \tag{20.8}$$

$$\mathcal{H}'|0_A 1_B 0_C\rangle = -\frac{\hbar}{2}(-\omega^B - J^{AB} - J^{BC})|0_A 1_B 0_C\rangle.$$

The difference of energies for two states in (20.8) is,

$$\Delta E = \hbar(\omega^B + J^{AB} + J^{BC}), \tag{20.9}$$

which corresponds to the frequency ω_{00}^B in (20.5). The expressions for ω_{ik}^A and ω_{ik}^C are analogous to those given by the formulas (20.5). For the spins at the ends of the chain we have different frequencies. For example, in (20.2) we suppose that the left edge spin is spin A. The eigenfrequencies associated with the inversion of this spin are,

$$\omega_0^A = \omega^A + J^{AB}, \quad \omega_1^A = \omega^A - J^{AB}, \tag{20.10}$$

where ω_i^A means that the neighbor spin (B) is in the state i. Because of contribution of the edge atoms, we have a total of 16 eigenfrequencies associated with the inversion of one spin. We can easily understand the appearance of these frequencies by considering that neighboring spins produce an effective magnetic field on a given spin. The effective field can increase or decrease the external field, depending on the orientation of the neighboring spins.

Chapter 21

Digital Gates in a Spin Chain

So far one can not experimentally operate on individual spins such as on an ion in the ion traps. The problem of manipulation of quantum states using a spin system is rather complicated, and was not investigated experimentally so far. We consider in this chapter only the digital states $|0\rangle$ and $|1\rangle$, without superpositions and entanglements. So, the phase of the states is not important for us.

The first question for a spin system is – How can one manipulate a given qubit? We can use a π-pulse to drive a spin from the state $|0\rangle$ to $|1\rangle$, and vice versa. But this pulse definitely will affect a number of spins. This problem was solved by Lloyd [35], who suggested a special sequence of π-pulses which provides the exchange of the states between the neighboring spins. For example, to realize the exchange of states between the neighboring spins A and B, one can use the following sequence of π-pulses,

$$\omega_{01}^A \omega_{11}^A \omega_{10}^B \omega_{11}^B \omega_{01}^A \omega_{11}^A, \tag{21.1}$$

where $\omega_{ik}^{A,B}$ indicates a π-pulse with the frequency ω_{ik}^A (or ω_{ik}^B), and the sequence of pulses follows from the left to the right, i.e. the first π-pulse is the pulse with the frequency ω_{01}^A. The action of the sequence (21.1) is shown in Tbl. 21.1. One can see that the first pair of pulses changes the states of A atoms which have the right neighbors in the excited state,

$\omega_{01}^A \omega_{11}^A$	AB	A^*B	AB^*	A^*B^*
$\omega_{10}^B \omega_{11}^B$	AB	A^*B	A^*B^*	AB^*
$\omega_{01}^A \omega_{11}^A$	AB	A^*B^*	A^*B	AB^*
	AB	AB^*	A^*B	A^*B^*

Table 21.1: Change of the initial states of two neighboring atoms, A and B under the influence of the sequence (21.1). The asterisk indicates that the atom is in the excited state.

independent of the states of the left (C) neighbors. The second pair of pulses changes the states of B spins which have their left neighbors in the excited state, independent of the state of the right neighbor. Finally, the third pair of pulses repeats the action of the first pair. As a result, one has an exchange of one bit of information between A and B spins, independent of the states of the left and the right C-neighbors. One can use a simple sequences of pulses, like (21.1), to load the information into a spin chain. Let us, for example, have 6 spins, $ABCABC$. We want to load the number 7 into this chain. So, we should get the state $ABCA^*B^*C^*$ from the ground state, $ABCABC$. Here, by A^*, B^*, and C^* we denote the excited states of the corresponding spins. (Recall that the phase of the state is not important here, so we do not use Dirac notation.) Next, we list the frequencies for the sequence of π-pulses and the corresponding change of states of the spins,

$$1. \quad \omega_0^C: \quad ABCABC \rightarrow ABCABC^*, \qquad (21.2)$$

$$2. \quad \omega_{01}^B: \quad ABCABC^* \rightarrow ABCAB^*C^*,$$

$$3. \quad \omega_{01}^A: \quad ABCAB^*C^* \rightarrow ABCA^*B^*C^*.$$

The realization of the digital CN-gate is obvious. For example, if one applies two π-pulses with the frequencies ω_{10}^A and ω_{11}^A, the spin A changes its state only if the left C-neighbor spin is in the excited state, independent of the state of the right neighbor, B. We also show that using a slightly modified sequence (21.1), namely,

$$\omega_{01}^A \omega_{11}^A \omega_{11}^B \omega_{01}^A \omega_{11}^A, \qquad (21.3)$$

one can realize the F-gate (Tbl. 9.6), which is universal for digital computation [35]. The sequence (21.3) differs from (21.1) by the absence of a pulse with the frequency ω_{10}^B. If in the complex of spins, ABC, the spin C is in the excited state, the result of the action of the sequence of pulses (21.3) coincides with the result of the action of the sequence of pulses (21.1). If the spin C is in the ground state, the pulse ω_{11}^B does not act on the spin B. In this case, the result of the action of the sequence of pulses (21.3) on the neighboring spins A and B, coincides with the result of the action on these spins of the sequence of pulses,

$$\omega_{01}^A \omega_{11}^A \omega_{01}^A \omega_{11}^A. \tag{21.4}$$

The sequence (21.4) does not change the state of the A spin. So, the sequence (21.3) changes the states of the neighboring spins, A and B, only if the spin C is in the excited state, providing the F-gate with the C-spin as a control bit.

Besides the chain of nuclear spins, the promising candidates for the quantum computation are a chain of electron spins with Ising interactions, placed in an external magnetic field [52]; a heteropolymer in which each unit possesses a long-lived excited state [35]; and quantum dots [53].

Chapter 22

Non-resonant Action of π-Pulses

The difference in frequencies does not provide a 100% selective excitation of a given resonant spin because of non-resonant effects in the radio frequency pulses. This effect can be studied by explicit numerical calculations. Let us consider, as an example, the CN-gate based on a system of two spins [54], $I = 1/2$, placed in a constant homogeneous magnetic field which is pointed in the positive z direction. We take into consideration the Ising interaction between two spins and the interaction of each of these spins with an electromagnetic field which rotates in (x, y) plane (see Chapter 12). The Hamiltonian of the system can be written in the form,

$$\mathcal{H} = -\hbar(\gamma_1 \vec{B}\vec{I}_1 + \gamma_2 \vec{B}\vec{I}_2 + 2J I_1^z I_2^z), \qquad (22.1)$$

where the two spins have different gyromagnetic ratios, γ_1 and γ_2. The spins are interacting with a circularly polarized transverse electromagnetic field of the frequency, ω, (12.20). In this case, the Hamiltonian (22.1) can be rewritten in the form (see (12.22) and (12.23)),

$$\mathcal{H}/\hbar = -\sum_{k=1}^{2}\left\{\omega_k I_k^z + \frac{1}{2}\Omega_k(e^{-i\omega t} I_k^- + e^{i\omega t} I_k^+)\right\} - 2J I_1^z I_2^z, \quad (22.2)$$

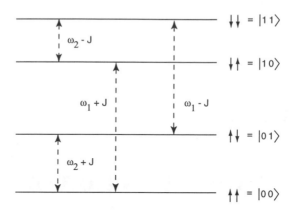

Figure 22.1: The energy levels of two Ising spins for $\omega_1 > \omega_2$. The dashed lines indicate single-spin transitions.

where $\omega_k = \gamma_k B^z$, $\Omega_k = \gamma_k h$ is the Rabi frequency (h is the dimensional amplitude of the transversal magnetic field). The energy levels of the two-spin system (22.1) are shown in Fig. 22.1, for the case, $\omega_1 > \omega_2$. If one applies a π-pulse with the frequency $\omega = \omega_2 - J$, it can provide a CN-gate with the first (left) spin as a control qubit, and the second (right) spin as a target qubit – the second spin changes its state only if the first spin is in the state $|1\rangle$.

Next, we present the results of numerical calculations of the dynamics of the CN-gate [54]. The time-dependent wave function of the Schrödinger equation (12.1), with the Hamiltonian (22.2), can be written as,

$$\Psi(t) = c_{00}(t)|00\rangle + c_{01}(t)|01\rangle + c_{10}(t)|10\rangle + c_{11}(t)|11\rangle. \quad (22.3)$$

We use a substitution which is equivalent to a transformation to the frame rotating with the frequency ω,

$$c_{00} \to c_{00}\exp(i\omega t + i\varphi(t)), \quad c_{01} \to c_{01}\exp(i\varphi(t)), \quad (22.4)$$

$$c_{10} \to c_{10}\exp(i\varphi(t)), \quad c_{11} \to c_{11}\exp(-i\omega t + i\varphi(t)),$$

where $\varphi(t)$ is a common phase which can be chosen arbitrarily to simplify the equations for the amplitudes c_{ik}. From the Schrödinger equation, we derive the equations for the amplitudes c_{ik},

$$-2i\dot{c}_{00}+2\dot{\varphi}c_{00}+2\omega c_{00} = \omega_1 c_{00}+\omega_2 c_{00}+J c_{00}+\Omega_1 c_{10}+\Omega_2 c_{01}, \quad (22.5)$$

$$-2i\dot{c}_{01} + 2\dot{\varphi}c_{01} = \omega_1 c_{01} - \omega_2 c_{01} - J c_{01} + \Omega_1 c_{11} + \Omega_2 c_{00},$$

$$-2i\dot{c}_{10} + 2\dot{\varphi}c_{10} = -\omega_1 c_{10} + \omega_2 c_{10} - J c_{10} + \Omega_1 c_{00} + \Omega_2 c_{11},$$

$$-2i\dot{c}_{11} + 2\dot{\varphi}c_{11} - 2\omega c_{11} = -\omega_1 c_{11} - \omega_2 c_{11} + J c_{11} + \Omega_1 c_{01} + \Omega_2 c_{10}.$$

The system of equations (22.5) was investigated numerically in [54]. The following values of parameters were chosen,

$$\omega_1 = 500, \quad \omega_2 = 100, \quad J = 5, \quad \omega = \omega_2 - J = 95, \quad (22.6)$$

$$\Omega_1 = 0.5, \quad \Omega_2 = 0.1.$$

The characteristic dimensional parameters, for the case of nuclear spins, can be obtained, for example, by multiplying the parameters in (22.6) by the factor $2\pi \times 10^6 s^{-1}$, which corresponds to the frequency 1 MHz. The condition, $\omega = \omega_2 - J = 95$, corresponds to the resonant transition: $|10\rangle \rightarrow |11\rangle$, i.e. the target qubit changes its state only if the control qubit is in the excited state. The phase $\varphi(t) = (\omega_2 - \omega_1 - J)t/2$ was chosen.

The dynamics of the modulus of the amplitudes $|c_{10}(t)|$ and $|c_{11}(t)|$ for the parameters (22.6), and for the initial conditions,

$$c_{10}(0) = 1, \quad c_{00} = c_{01}(0) = c_{11}(0) = 0. \quad (22.7)$$

is shown in Fig. 22.2. One can see that under the action of a π-pulse, the value of $|c_{10}|$ approaches zero, and the value of $|c_{11}|$ approaches 1. The two other amplitudes are zero. It means that the second spin transfers from the state $|0\rangle$ to the state $|1\rangle$, while the first spin remains in the same state, $|1\rangle$. Similarly, the state $|11\rangle$ transforms to the state $|10\rangle$. At the same time, a π-pulse does not affect the states $|00\rangle$ and $|01\rangle$. Thus, this system provides the digital CN-gate for quantum states in

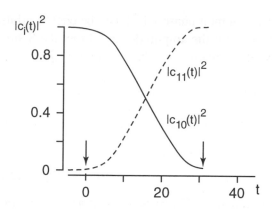

Figure 22.2: Time-dependence of $|c_{10}(t)|$ and $|c_{11}(t)|$. The vertical arrows indicate the beginning and the end of the action of the rectangular π-pulse.

spite of the non-resonant effects. (The dynamics of the quantum CN-gate will be discussed in Chapter 25.) As numerical experiments show, the angle of rotation of the target spin, α, is slightly bigger than $\Omega_2\tau$, where τ is the duration of the electromagnetic pulse. The reason for this is associated with the weak indirect excitation of the resonant transition via non-resonant spin. The terms $\Omega_1 c_{00}$ and $\Omega_1 c_{01}$ in the last two equations (22.5) are responsible for these effects.

The next question is how to avoid the effects of the non-resonant influence of the rotating magnetic field, when these effects are significant. Assume, for example, that one applies the electromagnetic pulse with the frequency ω_{01}^B in a chain of nuclear spins $ABCABC...$ (see Chapter 20). This pulse also influences any spin B which has the eigenfrequency ω_{10}^B, because these two frequencies only differ by a small value, $2(J^{AB} - J^{BC})$. Following [55], we estimate the influence of the non-resonant excitations. Let us consider the deflection of spin B with the frequency ω_{10}^B, under the action of an electromagnetic pulse with the frequency ω_{01}^B. We suppose that this deflection is small, and consider the influence of the neighbor spins, A^* and C, (spin A^* is in the excited state, and spin C is in the ground state). We use the language of the "effective field", which is often used in describing dynamics of magnetic systems [56]. Suppose that a spin A^*, which points "downward", produces the effective magnetic field on the neighboring spin B which is equal to

$$\vec{B}_1 = -\vec{e^z} J^{AB}/\gamma_B, \qquad (22.8)$$

where γ_B is the gyromagnetic ratio of spin B and $\vec{e^z}$ is the unit vector pointed in the positive z-direction. In the same way, the neighboring spin, C, which points "upward", produces the effective field which acts on the spin B,

$$\vec{B}_2 = \vec{e^z} J^{BC}/\gamma_B. \qquad (22.9)$$

So, before the action of a π-pulse, the spin B experiences the action of the net effective magnetic field of magnitude, B_e,

$$B_e = B_0 + (J^{BC} - J^{AB})/\gamma_B, \qquad (22.10)$$

which points in the positive z-direction. (B_0 is the external permanent magnetic field.) This field provides the frequency of transition between the states $|0\rangle$ and $|1\rangle$ for the considered spin B, which is equal to (see (12.3)),

$$\gamma_B B_e = \omega_{10}^B. \qquad (22.11)$$

Because the phase of the wave function is not important here, it is convenient to use the equations of motion for the average spin. To get these equations, we consider the spin dynamics in the Heisenberg representation. In this representation, the wave function is time independent, but the quantum-mechanical operators depend on time [48]. In the Heisenberg representation, the equation of motion for the vector spin operator, $\vec{I}$, is given by the Heisenberg equation,

$$i\hbar \frac{d}{dt}\vec{I} = [\vec{I}, \mathcal{H}], \qquad (22.12)$$

where $\mathcal{H}$ is the Hamiltonian, and $[\vec{I}, \mathcal{H}]$ is the commutator,

$$[\vec{I}, \mathcal{H}] = \vec{I}\mathcal{H} - \mathcal{H}\vec{I}. \qquad (22.13)$$

The Hamiltonian $\mathcal{H}$ is given by (12.21), where we replace the external field, $\vec{B}$, by the effective field, $\vec{B}_e$, and $\gamma = \gamma_B$. Using an implicit representation of the operator $\vec{I}$ (12.4) and (12.10), one can derive the well-known relations,

$$I^x I^y - I^y I^x = i I^z, \tag{22.14}$$

$$I^y I^z - I^z I^y = i I^x,$$

$$I^z I^x - I^x I^z = i I^y.$$

From (22.12)-(22.14) we derive the following equations,

$$\frac{d}{dt} I^x = \gamma_B (I^y B_e^z - I^z B_e^y), \tag{22.15}$$

$$\frac{d}{dt} I^y = \gamma_B (I^z B_e^x - I^x B_e^z),$$

$$\frac{d}{dt} I^z = \gamma_B (I^x B_e^y - I^y B_e^x).$$

Using the cross-product notation, we obtain the well-known equation which describes the dynamics of spin B,

$$\dot{\vec{I}} = \gamma_B \vec{I} \times \vec{B}_e. \tag{22.16}$$

Taking the quantum-mechanical average, we have the same equation for the average spin, $\langle \vec{I} \rangle$.

If one applies a circularly polarized electromagnetic pulse in the $x-y$ plane with the amplitude h and the frequency ω (see (12.20)), one has the effective magnetic field in (22.16), with the following components,

$$B_e^z = \omega_{10}^B / \gamma_B, \quad B_e^x = h \cos \omega t, \quad B_e^y = -h \sin \omega t. \tag{22.17}$$

Then, the equation for $\langle \vec{I} \rangle$ can be written in the form of the following three equations,

$$\frac{d}{dt} \langle I^+ \rangle + i \omega_{10}^B \langle I^+ \rangle = i \Omega_B e^{-i\omega t} \langle I^z \rangle, \tag{22.18}$$

$$\frac{d}{dt}\langle I^- \rangle - i\omega_{10}^B \langle I^- \rangle = -i\Omega_B e^{i\omega t} \langle I^z \rangle,$$

$$\frac{d}{dt}\langle I^z \rangle = \frac{i}{2}\Omega_B \left(\langle I^+ \rangle e^{i\omega t} - \langle I^- \rangle e^{-i\omega t} \right),$$

where $\Omega_B = \gamma_B h$, and $\langle I^\pm \rangle = \langle I^x \rangle \pm i \langle I^y \rangle$. The substitution,

$$\langle I^+ \rangle = s e^{-i\omega t}, \qquad \langle I^- \rangle = s^* e^{i\omega t}, \tag{22.19}$$

is equivalent to a transition to the rotating reference frame. Introducing the notation, $m = \langle I^z \rangle$, we finally derive from (22.18) the system of equations which describes the dynamics of the average spin,

$$\dot{s} + i(\omega_{10}^B - \omega)s = i\Omega_B m, \tag{22.20}$$

$$\dot{m} = \frac{i}{2}\Omega_B(s - s^*).$$

For a resonant pulse, $\omega = \omega_{10}^B$, and the initial conditions, $m(0) = 1/2$, $s(0) = 0$, we have the solution of (22.20),

$$s(t) = \frac{i}{2}\sin\Omega_B t, \tag{22.21}$$

$$m(t) = \frac{1}{2}\cos\Omega_B t.$$

We now take into consideration the fact that the real and imaginary parts of s describe the dynamics of the x and y – components of the average spin in the rotating frame. So, the solution (22.21) coincides with the expressions (12.37) derived directly from the wave function of the corresponding Schrödinger equation.

Consider now the non-resonant case, when the frequency of the electromagnetic field satisfies the following relation, $\omega = \omega_{01}^B \neq \omega_{10}^B$. So, the electromagnetic pulse is intended to excite a spin B, with the frequency ω_{01}. In this case, the solution of (22.20) can be written as,

$$s(t) = \pm\frac{1}{2}\sin\theta[2\cos\theta\sin^2(\omega_e t/2) + i\sin(\omega_e t)], \tag{22.22}$$

$$m(t) = \pm \frac{1}{2}[1 - 2\sin^2\theta \sin^2(\omega_e t/2)],$$

where the upper sign "+" corresponds to the initial conditions,

$$m(0) = \frac{1}{2}, \quad s(0) = 0, \tag{22.23}$$

and the lower sign "-" corresponds to the initial conditions,

$$m(0) = -\frac{1}{2}, \quad s(0) = 0, \tag{22.24}$$

and the following notation is introduced,

$$\sin\theta = \Omega_B/\omega_e, \quad \cos\theta = (\omega_{10}^B - \omega_{01}^B)/\omega_e, \tag{22.25}$$

$$\omega_e = \sqrt{\Omega_B^2 + (\omega_{10}^B - \omega_{01}^B)^2}.$$

Expressions (22.22) describe the precession of the non-resonant spin, B, in the rotating frame, around the effective field, $\vec{B}_e$, with the z-component, $(\omega_{10}^B - \omega_{01}^B)/\gamma_B$, and the x-component, Ω_B/γ_B. In (22.25), ω_e is the frequency of spin precession around the effective field, $\vec{B}_e$, and θ is the polar angle of the effective field. Fig. 22.3 shows the precession of the average spin, $\langle \vec{I} \rangle$ in the vicinity of the ground state $(m(0) = 1/2)$, for the case, $\omega_{10}^B \rangle \omega_{01}^B$. The deviation of the spin B from the digital states, $m = \pm(1/2)$ will be small only if the amplitude of the pulse, h, is small in comparison with the frequency difference, $|\omega_{10}^B - \omega_{01}^B|/\gamma_B$.

Now we shall show how one can eliminate the non-resonant deviation of the average spin. It is clear from (22.22), that a non-resonant spin returns to its initial position at the end of a π-pulse if $\omega_e \tau = 2\pi k$, $k = 1.2, ...$; τ is the duration of the π-pulse. Setting

$$\Omega_B \tau = \pi, \quad \omega_e \tau = 2\pi k, \quad k = 1, 2, ..., \tag{22.26}$$

one gets the π-pulse for the resonant spin which does not deflect the non-resonant spin. To realize such an opportunity, one should choose the amplitude and the duration of a pulse to satisfy the conditions,

$$\Omega_B = |\omega_{01}^B - \omega_{10}^B|/(4k^2 - 1)^{1/2}, \quad \tau = \pi/\Omega_B. \tag{22.27}$$

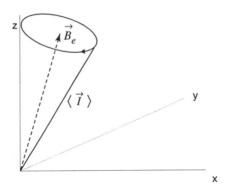

Figure 22.3: Precession of the non-resonant spin, $\langle \vec{I} \rangle$, around the effective field, $\vec{B}_e$, in the rotating frame.

This choice eliminates the non-resonant deviation of the spin B from both initial conditions, $m^{(0)} = 1/2$ and $m^{(0)} = -1/2$. The same analysis is valid for spins A and C.

Chapter 23

Experimental Logic Gates in Quantum Systems

A one-qubit rotation, under the action of a resonant π-pulse, is a commonly used experimental technique. That is why the present efforts of experimental groups are concentrated on the design of the two-qubit quantum logic gates. Recall that the quantum CN-gate, in combination with the one-qubit rotations, is universal for quantum computation, i.e. any logic gate can be constructed from their combinations [20]. The same is true for the CZ-gate and one-qubit rotations, because the CN-gate itself can be obtained as a combination of the CZ-gate and one-qubit rotations (see Chapter 18).

We now describe the first experimental realization of the CN-gate in a quantum system, demonstrated by Monroe et al. [22]. These authors used a modified Cirac and Zoller method for a single $^9Be^+$ ion in a *rf* ion trap. The target qubit was spanned by two hyperfine levels $|0\rangle = |F = 2, m_F = 2\rangle$ and $|1\rangle = |F = 1, m_F = 1\rangle$ of $^2S_{1/2}$ electronic ground state. Here F denotes the total (electron+nuclear) spin of the state, m_F denotes the projection of the total spin on the direction of the external magnetic field. The control qubit was spanned by the first two vibrational states of the trapped ion. The energy levels are shown in Fig. 23.1, where the first number k in $|kn\rangle$ belongs to the con-

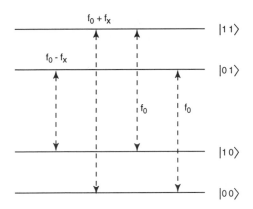

Figure 23.1: Energy levels used in the experiment [22]. Dashed lines show the main transitions used in this experiment.

trol qubit (vibrational state), the second number belongs to the target qubit (hyperfine state). The hyperfine frequency $f_0 = \omega_0/2\pi$ is approximately 1.25 GHz; the vibrational frequency $f_x = \omega_x/2\pi$ is approximately 11 MHz. In a weak magnetic field with $B_0 \approx 0.18$ mT, the levels $|0\rangle = |F = 2, m_F = 2\rangle$ and $|1\rangle = |F = 1, m_F = 1\rangle$ were separated from the lower Zeeman levels with $m_F < F$. To realize the quantum CN-gate the authors of [22] applied three pulses:

(1) a $\pi/2$-pulse for transitions $|k0\rangle \leftrightarrow |k1\rangle$ with a frequency f_0 ($k = 0, 1$).

(2) a 2π-pulse for the auxiliary transition between the state $|11\rangle$ and the level corresponding to the $|F = 2, m_F = 0\rangle$ and the ground vibrational state $|0\rangle$. (This state does not carry qubits and is not shown in Fig. 23.1. It is separated from the level $|00\rangle$ by approximately 2.5 MHz.) This transition reverses the sign of the state $|11\rangle$, namely: $|11\rangle \rightarrow -|11\rangle$.

(3) a $\pi/2$-pulse for the transition $|k0\rangle \leftrightarrow |k1\rangle$ with a frequency f_0 and a π-phase shift relative to the first $\pi/2$-pulse. For the transition $|00\rangle \leftrightarrow |01\rangle$, the effects of these two $\pi/2$ pulses cancel each other, and the ion remains in the initial state $|00\rangle$ or $|01\rangle$. But for the $|10\rangle \leftrightarrow |11\rangle$ transition, the effects of two $\pi/2$-pulses are additive resulting in the

transformation from one state to the other.

In the experiment [22], a single $^9Be^+$ ion was stored in a radio frequency ion trap with vibrational frequencies $f_x, f_y, f_z \approx 11, 18, 30$ MHz. Employing the stimulated Raman cooling in the x-dimension, the authors achieved a 95% time occupation in the vibrational ground state. To induce the transitions, they applied two Raman beams of approximately 1 milliwatt power at 313 nm which were detuned approximately by 50 GHz to the red from the $^2P_{1/2}$ excited state. The wave-vector difference pointed approximately along the x-axis of the trap, so the Raman transitions were insensitive to the y– and z-directions of motion. The difference of the frequencies of two Raman beams could be set to any of the three frequencies, f_0, $f_0 + f_x$, and $f_0 - f_x$; it was tunable from 1.2 to 1.3 GHz.

To detect the population of a target qubit, the authors applied σ_+ polarized laser radiation to the transitions $|k0\rangle \leftrightarrow {}^2P_{3/2}|F = 3, m_F = 3\rangle$. Then they detected the ion fluorescence which was proportional to the population of $|k0\rangle$ states. To detect the population of the control qubit, they added a Raman π-pulse. For example, if they got the value of a target qubit "0" ($|k0\rangle$ states), they could repeat the experiment applying a Raman π-pulse with a frequency $f_0 - f_x$ just prior to the detection of a target qubit. The presence of fluorescence indicated the value "0" for the control qubit.

To prepare any initial state of qubits, the authors applied one or two Raman π-pulses to the $|0, 0\rangle$ state. Then applying prescribed sequences of Raman pulses to any of the four initial states of qubits, they demonstrated a reliable implementation of the quantum CN-gate with digital initial conditions.

Another realistic idea for implementation of the CN-gate is the use of cavity quantum electrodynamics (QED) techniques [53, 57]. (Review of this technique are given in [58].) The state of the cavity field may be either a vacuum $|0\rangle$ or a single-photon state, $|1\rangle$. When an atom passes through the cavity, and the cavity is tuned to the atomic transition, the interaction Hamiltonian can be written as [57]:

$$\mathcal{H}_1 = i\hbar\Omega_1(|0\rangle\langle 1|a^\dagger - |1\rangle\langle 0|a). \tag{23.1}$$

Here, the states $|0\rangle$ and $|1\rangle$ refer to the atom, Ω_1 is the single-photon Rabi frequency; and $a^\dagger$, a are creation and annihilation operators of a photon in the cavity. If the cavity is detuned from the transition frequency, the non-resonant Hamiltonian can be written as [57],

$$\mathcal{H}_2 = \hbar(\Omega_2/2)a^\dagger a(|1\rangle\langle 1| - |0\rangle\langle 0|), \tag{23.2}$$

where $\hbar\Omega_2$ is the change of the atomic level spacing per photon in the cavity. It was suggested in [53, 57] that one uses a non-resonant interaction (23.2) to produce a phase shift in the atomic state controlled by the photon number. This method is related to Ramsey atomic interferometry [59].

Let, for example, the target qubit be an atom with two circular Rydberg states $|0\rangle$ or $|1\rangle$ [53]. The high-Q cavity is placed between two auxiliary microwave cavities in which the classical microwave field produces a $\pi/2$-rotation of the effective spin corresponding to a two-level atom. After an atom passes through the first microwave cavity, the state of the system $|nk\rangle$ $(n, k = 0, 1)$ transforms to,

$$|n0\rangle \rightarrow \frac{1}{\sqrt{2}}|n\rangle(|0\rangle + i|1\rangle), \tag{23.3}$$

$$|n1\rangle \rightarrow \frac{1}{\sqrt{2}}|n\rangle(|1\rangle + i|0\rangle),$$

where the first number n refers to the photon in the high-Q cavity, and the second number k refers to the atom. In the central high-Q cavity, a non-resonant (dispersive) interaction with the quantized field produces a phase shift. For any state $|nk\rangle$ one has

$$|n0\rangle \rightarrow \exp(-in\Theta/2)|n0\rangle, \tag{23.4}$$

$$|n1\rangle \rightarrow \exp(in\Theta/2)|n1\rangle,$$

where Θ is the phase shift per photon which can be tuned to be π. The phase of the classical field in the third cavity is shifted by π relative to the first one. Then, in the third cavity

$$|n0\rangle \rightarrow \frac{1}{\sqrt{2}}|n\rangle(|0\rangle - i|1\rangle), \tag{23.5}$$

$$|n1\rangle \rightarrow \frac{1}{\sqrt{2}}|n\rangle(|1\rangle - i|0\rangle).$$

If there is a photon in the high-Q cavity ($n = 1$), then the phase shift for an atom in this cavity is equal to π, and the rotations of an effective spin in the edge cavities add, producing a total rotation of π. In the opposite case, ($n = 0$) both rotations in the edge cavities cancel each other, and the atom remains in the initial state. The typical parameters of the system are: the resonant frequency $\approx 2 \times 10^{10}$ Hz, and the cavity field life-time $\approx 0.5s$ [53].

We shall describe now the measurement of a conditional phase shift (quantum phase gate) in a QED cavity, performed by Turchette et al. [24]. In this experiment, the qubits were spanned by two photons with different frequencies. Conditional dynamics originated from the nonlinear optical response of a cesium atom. Two atomic transitions with the orthogonal circular polarizations σ_+ and σ_- were coupled to the cavity field. The rate of the σ_- transition $6S_{1/2}$ $|F = 4, m = 4\rangle \leftrightarrow 6P_{3/2}$ $|F = 5, m = 3\rangle$ was negligible compared with the σ_+ transition to $|F = 5, m = 5\rangle$. The ground state $|F = 4, m = 4\rangle$ was prepared by optical pumping of a beam of Cs atoms. To study photon-photon interactions via an atom, the authors of [24] investigated the transmission of a pump beam with a frequency f_b, and a probe beam with a frequency f_a (Fig. 23.2). After these beams passed through the cavity, the polarization states of the beams were analyzed. If the probe beam was linearly polarized, then the σ_- component received a phase shift corresponding to an empty cavity. The σ_+ component received a shift corresponding to the atom-cavity system. The differential phase Φ_a between $\sigma_\pm$ components can be measured by analyzing the polarization of the output beam. To investigate the truth table for the quantum phase gate, the authors recorded the dependence of the probe beam phase Φ_a on the intensity of a pump field of either $\sigma_\pm$ polarization, and they did the same for the pump beam phase Φ_b. (The probe beam was detuned by 30 MHz and the pump beam was detuned by 20 MHz off the atomic resonance.) Then the authors extracted the phase shifts per photon to

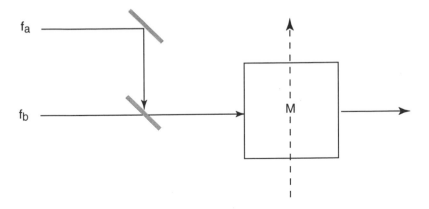

Figure 23.2: Scheme of the experiment [24]. Two photon beams with frequencies f_a and f_b interact in the cavity M via an atom. The dashed line shows the atomic beam.

get the following truth table:

$$|--\rangle \rightarrow |--\rangle, \tag{23.6}$$

$$|+-\rangle \rightarrow \exp(i\Phi_a)|+-\rangle,$$

$$|-+\rangle \rightarrow \exp(i\Phi_b)|-+\rangle,$$

$$|++\rangle \rightarrow \exp[i(\Phi_a + \Phi_b + \Delta)]|++\rangle,$$

where the first sign refers to the polarization of the probe beam, and the second refers to the polarization of the pump beam, $\Phi_a \approx 17.5^o$, $\Phi_b \approx 12.5^o$, $\Delta \approx 16^o$.

We shall discuss also very briefly the opportunities for using quantum dots to implement quantum logic gates. Barenco et al. [53] suggested using quantum dots to implement a quantum CN-gate. They considered two single-electron quantum dots embedded in a semiconductor and separated by a distance R. A qubit is spanned by the ground state $|0\rangle$ and the first excited state $|1\rangle$ of a dot. In the presence of an external static electric field, one can choose the reference frame in which the dipole moment in the state $|0\rangle$ is d_i and the dipole moment in the

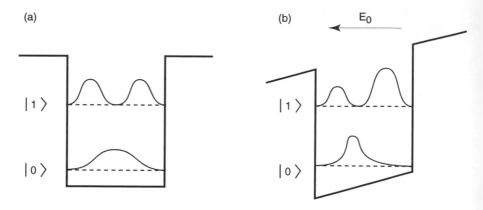

Figure 23.3: The wave functions of a quantum dot for electric field $E_0 = 0$ (a); and $E_0 \neq 0$ (b) [53].

state $|1\rangle$ is $-d_i$ $(i = 1, 2)$. The observable effect is connected with the shift of the charge distribution which is opposite in the states $|0\rangle$ and $|1\rangle$ (Fig. 23.3). It is supposed that the two dots, A and B, have different resonant frequencies ω^A and ω^B. The Hamiltonian of the dipole-dipole interaction $\mathcal{H}_{int}$ to a good approximation commutes with the Hamiltonian of noninteracting dots, and,

$$\mathcal{H}_{int}|nk\rangle = (-1)^{n+k+1}\hbar(J/2)|nk\rangle, \quad J = \frac{d_1 d_2}{2\pi \varepsilon_0 R^3}, \quad (23.7)$$

where $n = 0, 1$ refers to the state of the first dot, and $k = 0, 1$ refers to the state of the second dot. Thus, the dipole-dipole interaction between the dots can be described by the Ising interaction (see (20.1)) of two effective spins with the constant of interaction J. The scheme for energy levels of the two-dot system is the same as the energy levels of the two-spin system (Fig. 22.1). To realize the quantum CN-gate with the control qubit A, it was suggested in [53] to apply a π-pulse with the frequency $\omega^B - J$. This pulse drives the dot B if the dot A is in the excited state.

Chapter 24

Error Correction for Quantum Computers

One of the main challenges for a theory of quantum computers is the problem of error correction. The standard way to correct a computer error uses redundancy. One uses several elements to represent the same bit. To incorporate the error correction procedure into a digital atomic chain computer, Lloyd suggested using the more complicated three-level systems [18]. The atoms (A, B, or C) must possess an additional excited state $|2\rangle$ which decays quickly into the ground state.

Let, for example, the atom B have a state $|2\rangle$ which decays quickly into the ground state $|0\rangle$. Assume, that a triplet ABC is used to store the same bit, i.e., one has the state $|000\rangle$ or $|111\rangle$ (here $|ijk\rangle$ means $|i_A j_B k_C\rangle$). An error usually changes the state of only one atom. Then, one would have one of the states with an error $|001\rangle$, $|010\rangle$, or $|100\rangle$, instead of the state $|000\rangle$ (or the states $|011\rangle$, $|101\rangle$, or $|110\rangle$ instead of $|111\rangle$).

To correct the error, one applies the sequence of pulses with the frequencies,

$$\omega_{00}^B(1 \leftrightarrow 2)\omega_{11}^B(0 \leftrightarrow 1)\omega_{11}^B(1 \leftrightarrow 2)\omega_{11}^B(0 \leftrightarrow 1), \qquad (24.1)$$

where $\omega_{ik}^B(n \leftrightarrow m)$ denotes the frequency of transition $n \leftrightarrow m$ ($n, m =$

0, 1, 2) for atom B when the neighboring atoms A and C are in the states $|i\rangle$ and $|k\rangle$ (the first pulse has the frequency ω_{00}^B). Then, one exchanges the states of A and B (see Chapter 21), and repeats the sequence (24.1). At the third step, one exchanges the states of the atoms B and C, and again repeats (24.1). As a result, the initial state $|000\rangle$ or $|111\rangle$ is restored.

As an example, we consider the correction of the state $|001\rangle$. At the first step, the sequence (24.1) does not change this state because it acts on the atom B only if the both neighbors (A and C) are in the same state, either in $|0\rangle$ or in $|1\rangle$. After the atoms A and B exchange their states, one has the same situation (the state $|001\rangle$). After the exchange of states between the atoms B and C, one has the state $|010\rangle$. Now, the first pulse of the sequence (24.1) drives the atom B to the state $|2\rangle$, (see Fig. 24.1a) which quickly decays to the state $|0\rangle$. The three other pulses do not act on the atom B. As a result, one has the desired state $|000\rangle$. In Fig. 24.1b, we show the analogous scheme for the case when the neighbors, A and C, are in the excited state (the state $|101\rangle$). We shall discuss now important objections which concern the action of the electromagnetic pulses [60]. Let consider first a chain of spins $ABCABC...$ with the Ising interaction, and an interaction with a resonant electromagnetic π-pulse. It is obvious, that the angle of rotation, α_i, may occasionally differ from π, where,

$$\alpha_i \approx \gamma_i h\tau, \quad (i = A, B, C), \tag{24.2}$$

where h is the amplitude, and τ is the duration of the electromagnetic pulse which is circularly polarized in the xy plane. (We mentioned already in Chapter 22 that the angle of rotation, α, is slightly larger than $\gamma_i h\tau$.)

Now we consider one of the opportunities to correct the error caused by the distortion of the π-pulse in a spin chain [55]. Assume that the electromagnetic radiation enters the quantum computer through one of 16 resonant samples, where the signal of the spin resonance could be observed. Three samples contain the spins A, three samples – spin B, and three samples – spin C. The extra four samples correspond to the

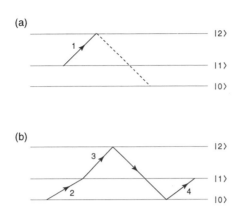

Figure 24.1: Correction of the "error bit" in atom B in the triplet ABC; (a) The neighbors A and C are in the ground state $|010\rangle$; (b) The neighbors A and C are in the excited state $|101\rangle$. The numbers $i = 1 - 4$ denote the transition under the action of the i-th pulse in the sequence (24.1).

edge spins.

The scheme for correction of the distorted π-pulses for the "spin quantum computer" is presented in Fig. 24.2. Every sample "S_j" is placed in the magnetic field $\vec{B}_j$ oriented along the z axis which corresponds to one of the resonant frequencies of the quantum computer (12 internal frequencies, and four frequencies for the edge ions). For example, for the samples with the spins B,

$$\gamma_B B_1 = \omega_{00}^B, \quad \gamma_B B_2 = \omega_{01}^B, \quad \gamma_B B_3 = \omega_{10}^B, \quad \gamma_B B_4 = \omega_{11}^B,$$

where B_1, B_2, and B_3 are the corresponding magnitudes of the magnetic field. The amplitudes $h_{A,B,C}$ of the electromagnetic pulses are selected in such a way, that $\alpha_{A,B,C} = \pi$. So, for the normal situation (undistorted π-pulses) the signal of the free precession (SFP) from the "resonant sample" is absent. If occasionally $\alpha_B \neq \pi$, the SFP from the resonant sample may be measured.

Let us assume, for example, that the electromagnetic π-pulse drives a spin B from the ground state, $|0\rangle$, to the excited state, $|1\rangle$. The evo-

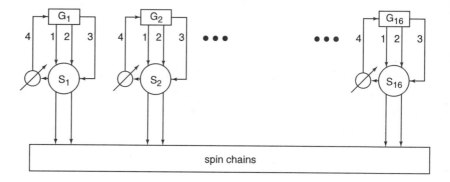

Figure 24.2: The scheme for correction of the distorted π-pulses. The boxes G_{1-16} are the "generators" of π-pulses and "weak correcting pulses". The circles S_{1-16} are the "resonant samples". Ø are the measuring devices which measure the signal of the free precession and transfer the information (lines "4") to the "generators", to create a weak correcting pulse. "1" - the π-pulses, "2" - "weak correcting pulses", "3" - restored pulses, which act only on the resonant samples, to restore the equilibrium state.

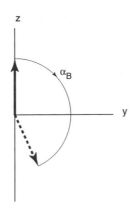

Figure 24.3: Rotation of spin B around the x-axis of the rotating frame under the action of the distorted π-pulse.

lution of the average spin can be described approximately by the solution (22.21). Suppose, for example, one detects the SFP after the action of a π-pulse, $s = ia$, in a resonant sample. If the measured value, a, is greater than zero, then the y-component of the average spin ($\langle I^y \rangle = Im(s)$) in the rotating system of coordinates is positive. It means that the angle of rotation, $\alpha_B < \pi$ (see Fig. 24.3). In this case, one applies a resonant pulse along the x-axis of the rotating system of coordinates, with the angle of rotation, α'_B,

$$\alpha'_B = \pi - \alpha_B = \arcsin 2a. \tag{24.3}$$

If $a < 0$, it means that $\alpha_B > \pi$. In this case, one should apply an additional pulse with $\alpha'_B = \alpha_B - \pi = \arcsin |2a|$, along the negative X-axis of the rotating system of coordinates, i.e. a pulse with a phase shift π relative to the first pulse. Note, that after the additional pulse, one should apply a π-pulse to the sample to restore the initial state of the resonant sample.

One of the most challenging problems in quantum computation is the error correction for complicated superpositional states which are used, in particular, in Shor's algorithm. Discovery of the existence of error correction codes for the superpositional quantum states [25, 61, 62] was the second triumph of quantum computation theory, after the discovery of the prime factorization algorithm. We shall describe here the principal opportunity to correct the phase error of a spin system based on the simple three-qubit scheme suggested by Steane [61].

To describe this scheme, we make a transformation to the I^x-representation, in which the matrix I^x is diagonal. This transformation can be done using the unitary matrix,

$$U = \frac{1}{\sqrt{2}} \begin{pmatrix} 1 & -1 \\ 1 & 1 \end{pmatrix}, \qquad (24.4)$$

and formulas (15.5). One can check that,

$$(I^x)' = U^\dagger I^x U = \qquad (24.5)$$

$$\frac{1}{4} \begin{pmatrix} 1 & 1 \\ -1 & 1 \end{pmatrix} \begin{pmatrix} 0 & 1 \\ 1 & 0 \end{pmatrix} \begin{pmatrix} 1 & -1 \\ 1 & 1 \end{pmatrix} = \frac{1}{2} \begin{pmatrix} 1 & 0 \\ 0 & -1 \end{pmatrix}$$

Thus, in the I^x-representation, the I^x matrix is diagonal, and its eigenfunctions,

$$|0\rangle^x = \begin{pmatrix} 1 \\ 0 \end{pmatrix}_x, \qquad |1\rangle^x = \begin{pmatrix} 0 \\ 1 \end{pmatrix}_x, \qquad (24.6)$$

correspond to the eigenvalues, $I^x = \pm 1/2$. We shall use the CN-gate in the x-representation,

$$\text{CN}_{ik}^x = |0_i 0_k\rangle\langle 0_i 0_k|^x + |0_i 1_k\rangle\langle 0_i 1_k|^x + \qquad (24.7)$$

$$|1_i 0_k\rangle\langle 1_i 1_k|^x + |1_i 1_k\rangle\langle 1_i 0_k|^x,$$

which transfers the state of the target spin, k, if the control spin, i, points in the negative x-direction (the state $|1_i\rangle^x$). Assume that the initial qubit is in an arbitrary state, which can be described by the wave function,

$$\psi_0 = c_0|0_1\rangle + c_1|1_1\rangle. \qquad (24.8)$$

Here, we omit the normalization constant, which is not important. For encoding this superpositional state, we consider two additional qubits in the states, $|0_2\rangle$ and $|0_3\rangle$. Then, we apply the operators CN_{12}^x and CN_{13}^x to the wave function,

$$\Psi_0 = (c_0|0_1\rangle + |c_1|1_1\rangle)|0_2 0_3\rangle = \qquad (24.9)$$

$$c_0|0_1 0_2 0_3\rangle + c_1|1_1 0_2 0_3\rangle.$$

As a result, we obtain the encoded entangled state,

$$\Psi_1 = \mathrm{CN}_{13}^x \mathrm{CN}_{12}^x \Psi_0 = c_0|0_1 0_2 0_3\rangle + c_1|1_1 1_2 1_3\rangle. \qquad (24.10)$$

Now we consider a phase distortion for one of three qubits, for example, for the first one. We represent this distortion in the form,

$$|0_1\rangle \rightarrow |0_1\rangle \cos(\varphi/2) + i|1_1\rangle \sin(\varphi/2), \qquad (24.11)$$

$$|1_1\rangle \rightarrow |1_1\rangle \cos(\varphi/2) + i|0_1\rangle \sin(\varphi/2).$$

To confirm that (24.11) describes the phase distortion, let us compare the state $c_0|0\rangle + c_1|1\rangle$ with the distorted state,

$$c_0\{\cos(\varphi/2)|0\rangle + i\,\sin(\varphi/2)|1\rangle\}+ \qquad (24.12)$$

$$c_1\{\cos(\varphi/2)|1\rangle + i\,\sin(\varphi/2)|0\rangle\} =$$

$$\{c_0 \cos(\varphi/2) + i c_1 \sin(\varphi/2)\}|0\rangle+$$

$$\{c_1 \cos(\varphi/2) + i c_0 \sin(\varphi/2)\}|1\rangle.$$

We are going to compare the averages, $\langle I^x\rangle$, $\langle I^y\rangle$, and $\langle I^z\rangle$ for these two functions. In the I^x-representation, the I^z matrix can be found using (15.5) and (24.4),

$$(I^z)' = \frac{1}{4}\begin{pmatrix} 1 & 1 \\ -1 & 1 \end{pmatrix}\begin{pmatrix} 1 & 0 \\ 0 & -1 \end{pmatrix}\begin{pmatrix} 1 & -1 \\ 1 & 1 \end{pmatrix} = -\frac{1}{2}\begin{pmatrix} 0 & 1 \\ 1 & 0 \end{pmatrix} \quad (24.13)$$

In the same way, we can find the matrix I^y in the I^x-representation,

$$(I^y)' = \frac{1}{4}\begin{pmatrix} 1 & 1 \\ -1 & 1 \end{pmatrix}\begin{pmatrix} 0 & -i \\ i & 0 \end{pmatrix}\begin{pmatrix} 1 & -1 \\ 1 & 1 \end{pmatrix} = \frac{1}{2}\begin{pmatrix} 0 & -i \\ i & 0 \end{pmatrix},$$

$$(24.14)$$

i.e. the matrix I^y is the same in both representations. Now we can find the average values, $\langle I^x \rangle$, $\langle I^y \rangle$ and $\langle I^z \rangle$.

For the initial state, $c_0|0\rangle + c_1|1\rangle$, we have,

$$\langle I^x \rangle_i = \frac{1}{2}(c_0^* c_1^*) \begin{pmatrix} 1 & 0 \\ 0 & -1 \end{pmatrix} \begin{pmatrix} c_0 \\ c_1 \end{pmatrix} = \frac{1}{2}(|c_0|^2 - |c_1|^2), \qquad (24.15)$$

$$\langle I^y \rangle_i = \frac{1}{2}(c_0^* c_1^*) \begin{pmatrix} 0 & -i \\ i & 0 \end{pmatrix} \begin{pmatrix} c_0 \\ c_1 \end{pmatrix} = \frac{i}{2}(c_0 c_1^* - c_0^* c_1),$$

$$\langle I^z \rangle_i = -\frac{1}{2}(c_0^* c_1^*) \begin{pmatrix} 0 & 1 \\ 1 & 0 \end{pmatrix} \begin{pmatrix} c_0 \\ c_1 \end{pmatrix} = -\frac{1}{2}(c_0 c_1^* + c_0^* c_1),$$

where index "i" means "initial". We used here the matrix representation of the wave function,

$$c_0|0\rangle + c_1|1\rangle = \begin{pmatrix} c_0 \\ c_1 \end{pmatrix}, \qquad (24.16)$$

$$c_0^* \langle 0| + c_1^* \langle 1| = (c_0^*, c_1^*),$$

and omitted "prime" for I^x, I^y, and I^z. In the same way, for the distorted wave function (24.12), we obtain,

$$\langle I^x \rangle_d = \frac{1}{2}[(|c_0|^2 - |c_1|^2) \cos\varphi + i\sin\varphi(c_1 c_0^* - c_0 c_1^*)] = \qquad (24.17)$$

$$\langle I^x \rangle_i \cos\varphi - \langle I^y \rangle_i \sin\varphi,$$

$$\langle I^y \rangle_d = \frac{1}{2}[i(c_0 c_1^* - c_1 c_0^*) \cos\varphi + (|c_0|^2 - |c_1|^2) \sin\varphi] =$$

$$\langle I^y \rangle_i \cos\varphi + \langle I^x \rangle_i \sin\varphi,$$

$$\langle I^z \rangle_d = \langle I^z \rangle_i,$$

where the index "d" indicates "distorted". Obviously, the expressions (24.17) describe the rotation of the vector $\langle \vec{I} \rangle$ in the (x, y) plane, by the angle φ, i.e. the phase distortion of the initial state.

After the distortion (24.11), the wave function Ψ_1 transforms to Ψ_2,

$$\Psi_2 = c_0[\cos(\varphi/2)|0_1 0_2 0_3\rangle + i\sin(\varphi/2)|1_1 0_2 0_3\rangle] + \qquad (24.18)$$

u_i	$|0_10_20_3\rangle$	$|0_10_21_3\rangle$	$|0_11_20_3\rangle$	$|1_10_20_3\rangle$
$(CN)^x_{12}u_i$	$|0_10_20_3\rangle$	$|0_10_21_3\rangle$	$|0_11_20_3\rangle$	$|1_11_20_3\rangle$
$(CN)^x_{13}(CN)^x_{12}u_i$	$|0_10_20_3\rangle$	$|0_10_21_3\rangle$	$|0_11_20_3\rangle$	$|1_11_21_3\rangle$

u_i	$|1_11_21_3\rangle$	$|1_11_20_3\rangle$	$|1_10_21_3\rangle$	$|0_11_21_3\rangle$
$(CN)^x_{12}u_i$	$|1_10_21_3\rangle$	$|1_10_20_3\rangle$	$|1_11_21_3\rangle$	$|0_11_21_3\rangle$
$(CN)^x_{13}(CN)^x_{12}u_i$	$|1_10_20_3\rangle$	$|1_10_21_3\rangle$	$|1_11_20_3\rangle$	$|0_11_21_3\rangle$

Table 24.1: The transformation of all possible terms u_i of the distorted wave function Ψ_2 under the action of two $(CN)^x$ operators.

$$c_1[\cos(\varphi/2)|1_11_21_3\rangle + i\,\sin(\varphi/2)|0_11_21_3\rangle].$$

Analogously, in the case of distortion of the second qubit by the phase φ, we get additional terms in Ψ_2 which are proportional to $|0_11_20_3\rangle$ and $|1_10_21_3\rangle$. Finally, for the distortion of the third qubit, we get additional terms in Ψ_2 which are proportional to $|0_10_21_3\rangle$ and $|1_11_20_3\rangle$.

For error correction, one applies to the state Ψ_2 the operators, CN^x_{12} and CN^x_{13}. Tbl. 24.1 shows the transformation of all possible terms (u_i in the Tbl. 24.1) of the distorted wave function $\Psi^{(2)}$. Now, one measures the x-components of the second and the third spins. If the measurement gives the result $I^x_2 = I^x_3 = -1/2$, then the "NOT-operator" $N^x_1 = |0_1\rangle\langle1_1|^x + |1_1\rangle\langle0_1|^x$ should be applied to the first qubit. (The operator N changes the state in the I^x representation.) For other outcomes of the measurement,

$$(I^x_2, I^x_3) = (1/2, -1/2), \quad (-1/2, 1/2), \quad (1/2, 1/2), \qquad (24.19)$$

the N^x operator is not applied. As a result, for any outcome one gets the initial undistorted wave function ψ_0 for the first spin.

For example, for the phase distortion of the first spin (formula (24.18)), we have after two CN^x operations,

$$\Psi_3 = c_0[\cos(\phi/2)|0_10_20_3\rangle + i\,\sin(\phi/2)|1_11_21_3\rangle]+ \qquad (24.20)$$

$$c_1[\cos(\phi/2)|1_1 0_2 0_3\rangle + i\sin(\phi/2)|0_1 1_2 1_3\rangle].$$

After measuring the second and the third qubits, we can get two outcomes,

$$1)I_2^x = I_3^x = 1/2, \quad 2)I_2^x = I_3^x = -1/2. \tag{24.21}$$

In the first case, we have for the wave function of the first spin,

$$c_0\cos(\phi/2)|0_1\rangle + c_1\cos(\phi/2)|1_1\rangle. \tag{24.22}$$

Omitting the insignificant common factor $\cos(\phi/2)$, we get from (24.22) the initial wave function ψ_0.

For the second case in (24.21), we have after a measurement, the wave function for the first spin,

$$ic_0\sin(\phi/2)|1_1\rangle + ic_1\sin(\phi/2)|0_1\rangle. \tag{24.23}$$

In this case, one should apply the N^x-operator to the wave function (24.23). Then, we have,

$$ic_0\sin(\phi/2)|0_1\rangle + ic_1\sin(\phi/2)|1_1\rangle. \tag{24.24}$$

Ignoring the common factor $i\sin(\phi/2)$ in (24.24), we again restore the initial wave function ψ_0.

According to the scheme suggested in [61], each qubit is encoded with three qubits, and then these qubits are corrected, and the initial qubit ψ_0 is restored. The decoding qubit can be used for computation, and then it is encoded again (see Fig. 24.4).

The physical implementation of such schemes is not simple, because we have to apply the CN-gate in the I^x-representation. Additional results and sugestions on quantum error correction can be found in the paper by J. P. Paz and W. H. Zurek (in page 355), and the paper by E. Knill R. Laflamme, W. H. Zurek (in page 365) in reference [41].

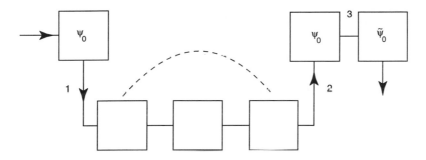

Figure 24.4: Steane's scheme for error correction; (1)-encoding of the initial qubit ψ_0 with the three qubits; (2)-error correction and decoding of the initial qubit; (3)-computation; $\tilde{\psi}_0$ is the initial qubit ψ_0 after computation.

Chapter 25

Quantum Gates in a Two-Spin System

We shall consider here a quantum two-qubit gate in the simplest system which contains only two spins with the Ising interaction. The energy levels of the system are shown in Fig. 22.1. The Hamiltonian of the system, including the interaction with the electromagnetic field, is given by the expression (22.2).

We shall discuss the opportunity of implementation of the quantum CN-gate using this two-spin system. Without an external electromagnetic field, the evolution of the system can be described by the following expression for the wave function, $\Psi(t)$,

$$\Psi(t) = \sum_{i,k=0}^{1} c_{ik}(0)e^{-iE_{ik}t/\hbar}|ik\rangle, \qquad (25.1)$$

where E_{ik} is the energy of the corresponding states,

$$E_{00} = -\frac{\hbar}{2}(\omega_1 + \omega_2 + J), \quad E_{01} = \frac{\hbar}{2}(-\omega_1 + \omega_2 + J), \qquad (25.2)$$

$$E_{10} = \frac{\hbar}{2}(\omega_1 - \omega_2 + J), \quad E_{11} = \frac{\hbar}{2}(\omega_1 + \omega_2 - J).$$

If one applies the electromagnetic pulses, then the wave function can be represented in the form (22.3). One can transform it to the "interaction representation",

$$c_{ik}(t) \rightarrow c_{ik}(t)\exp(-iE_{ik}t/\hbar), \tag{25.3}$$

which allows one to get rid of the phase factors corresponding to free evolution. Then one gets the equations for the amplitudes, $c_{ik}(t)$, which can be written in the form,

$$i\hbar\dot{c}_{ik} = \sum_{n,m}\langle ik|V|nm\rangle e^{i(E_{ik}-E_{nm})t/\hbar}c_{nm}, \tag{25.4}$$

where the Hamiltonian V describes the interaction between the two-spin system and the electromagnetic pulse,

$$V = -\frac{1}{2}\sum_{k=1}^{2}\Omega_k\left(e^{-i\omega t}I_k^{-} + e^{i\omega t}I_k^{+}\right). \tag{25.5}$$

However, equations (25.4) include rapidly oscillating time dependent coefficients, which makes them inconvenient for accurate numerical calculations of the dynamical behavior of the system.

We shall use another approach based on equations (22.5) which are written in the system of coordinates connected with the rotating magnetic field. Using the same substitution as in Chapter 22,

$$\omega = \omega_2 - J, \quad \varphi(t) = (\omega_2 - \omega_1 - J)t/2,$$

we derive the equations with time-independent coefficients,

$$-2i\dot{c}_{00} + 2(\omega_2 - \omega_1 - 2J)c_{00} = \Omega_1 c_{10} + \Omega_2 c_{01}, \tag{25.6a}$$

$$-2i\dot{c}_{01} + 2(\omega_2 - \omega_1)c_{01} = \Omega_1 c_{11} + \Omega_2 c_{00},$$

$$-2i\dot{c}_{10} = \Omega_1 c_{00} + \Omega_2 c_{11},$$

$$-2i\dot{c}_{11} = \Omega_1 c_{01} + \Omega_2 c_{10}.$$

Let us consider the free evolution of the two-spin system in the rotating frame. Substituting $\Omega_1 = \Omega_2 = 0$ in (25.6), we obtain the solution,

$$c_{00}(t) = c_{00}(0)e^{-i(\omega_2 - \omega_1 - 2J)t}, \qquad (25.6b)$$

$$c_{01}(t) = c_{01}(0)e^{-i(\omega_2 - \omega_1)t},$$

$$c_{10}(t) = c_{10}(0), \qquad c_{11}(t) = c_{11}(0).$$

To get rid of the phase factor corresponding to the free evolution, we will discuss instead of dynamics of the coefficients $c_{ik}(t)$, the dynamics of the coefficients,

$$c'_{00}(t) = c_{00}(t)e^{i(\omega_2 - \omega_1 - 2J)t}, \qquad (25.6c)$$

$$c'_{01}(t) = c_{01}(t)e^{i(\omega_2 - \omega_1)t}.$$

$$c'_{10}(t) = c_{10}(t), \qquad c'_{11}(t) = c_{11}(t).$$

The detailed investigations of the amplitudes, $c'_{ik}(t)$, (including their phases), under the action of the π-pulse, were carried out in [63]. In Fig. 25.1a, the time dependence of the real part, $Rec'_{11}(t)$, and the imaginary part, $Imc'_{10}(t)$ are shown, for the values of parameters (22.6), and initial conditions,

$$c'_{11}(0) = 1, \qquad c'_{ik}(0) = 0, \qquad (i, k) \neq (1, 1). \qquad (25.7)$$

One can see the monotonic decrease of $Rec'_{11}(t)$ and the increase of $Imc'_{10}(t)$, which describe the transition $|11\rangle \rightarrow \exp(i\pi/2)|10\rangle$. The values of $Rec'_{10}(t)$ and $Imc'_{11}(t)$ are negligible as are the values $|c'_{00}(t)|$ and $|c'_{01}(t)|$. In Fig. 25.1b, the analogous dependences are shown for the same values of parameters (22.6), and for the initial conditions,

$$c'_{10} = 1, \qquad c'_{ik} = 0, \qquad (i, k) \neq (1, 0). \qquad (25.8)$$

Now let us consider the initial conditions,

$$c'_{01} = 1, \qquad c'_{ik} = 0, (i, k) \neq (0, 1), \qquad (25.9)$$

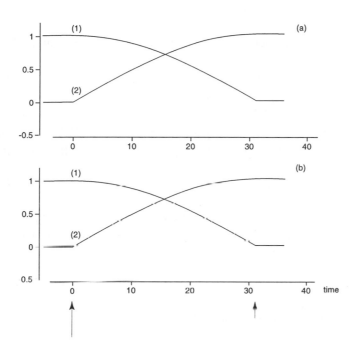

Figure 25.1: Time-evolution of the amplitudes c'_{ik} under the action of a π-pulse, for the initial conditions, (a) (25.7); (b) (25.8). In (a): curve (1) corresponds to the $Rec'_{11}(t)$, curve (2) corresponds to the $Imc'_{10}(t)$. In (b): curve (1) corresponds to the $Rec'_{10}(t)$, curve (2) corresponds to the $Imc'_{11}(t)$. The vertical arrows show the beginning and the end of the pulse.

which correspond to the population of the non-resonant level, $|01\rangle$. For these initial conditions, the amplitudes c'_{ik} practically do not change under the action of a π-pulse. The same is true for the initial conditions,

$$c'_{00}(0) = 1, \quad c'_{ik}(0) = 0, \quad (i, k) \neq (0, 0), \tag{25.10}$$

In Fig. 25.2 the action of the π-pulse on the superpositional initial state,

$$c'_{00}(0) = (0.3)^{1/2}, \quad c'_{01}(0) = 5^{-1/2}, \tag{25.11}$$

$$c'_{10}(0) = 3^{-1/2}, \quad c'_{11}(0) = 6^{-1/2},$$

is demonstrated, for the same parameters as in Fig. 25.1. One can see that at the end of the π-pulse the amplitudes take the following values,

$$c'_{11} = e^{i\pi/2}c'_{10}(0), \quad c'_{10} = e^{i\pi/2}c'_{11}(0). \tag{25.12}$$

The values of "non-resonant" amplitudes change little. Thus, the action of a π-pulse with the frequency $\omega_2 - J$ corresponds to the action of a two-qubit quantum gate,

$$|00\rangle\langle00| + |01\rangle\langle01| + e^{i\pi/2}|10\rangle\langle11| + e^{i\pi/2}|11\rangle\langle10|, \tag{25.13}$$

which can be considered as a modified CN-gate.

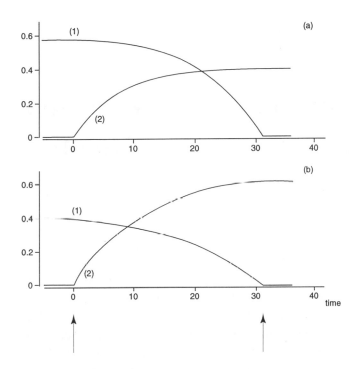

Figure 25.2: Time dependence of the amplitudes c'_{10} (a), and c'_{11} (b), for the superpositional initial conditions (25.11). In (a) the curve (1) corresponds to the $Rec'_{10}(t)$, the curve (2) corresponds to the $Imc'_{10}(t)$. In (b) the curve (1) corresponds to the $Rec'_{11}(t)$, the curve (2) corresponds to the $Imc'_{11}(t)$. The vertical arrows show the beginning and the end of the π-pulse.

Chapter 26

Quantum Logic Gates in a Spin Ensemble at Room Temperature

This chapter is based on the idea suggested independently in [28, 29] and [30], for quantum computation at room temperature. To explain the idea, let us first consider an ensemble of noninteracting spins, $I = 1/2$, in an external magnetic field which points in the positive z-direction. In the state of the thermal equilibrium, the system can be described by the density matrix (16.17) which for the typical condition, $k_B T \gg \hbar \omega_0$ has the form (16.19). The expression (16.19) consists of two terms: The first term, which corresponds to the infinite temperature, is proportional to the unit matrix, $\rho_\infty = (1/2)E$. This term does not influence the average spin, $\langle \vec{I} \rangle$, which can be measured experimentally for an ensemble of spins. Indeed, for example,

$$\langle I^x \rangle = \frac{1}{2} Tr\{I^x E\} = \frac{1}{2} Tr\{I^x\} = 0, \qquad (26.1)$$

as $I^x E = I^x$, and $Tr\{I^x\} = 0$. The same is true for the operators I^y and I^z.

The second term in (16.19) describes the deviation from ρ_∞,

$$\rho_\Delta = (\beta/4) \begin{pmatrix} 1 & 0 \\ 0 & -1 \end{pmatrix}, \quad \beta = \hbar\omega_0/k_B T. \tag{26.2}$$

The matrix ρ_Δ can be, in turn, represented as a sum of a diagonal matrix, ρ_a, which is proportional to E,

$$\rho_a = (\beta/4) \begin{pmatrix} -1 & 0 \\ 0 & -1 \end{pmatrix}, \tag{26.3}$$

and the matrix ρ_b,

$$\rho_b = (\beta/4) \begin{pmatrix} 2 & 0 \\ 0 & 0 \end{pmatrix} = (\beta/2) \begin{pmatrix} 1 & 0 \\ 0 & 0 \end{pmatrix}. \tag{26.4}$$

Again, the matrix ρ_a does not influence the average spin, $\langle \vec{I} \rangle$.

Now let us consider the evolution of the spin ensemble. This evolution can be caused, for example, by the application of resonant electromagnetic pulses. We shall present the time-dependent density matrix in the form,

$$\rho(t) = \frac{1}{2}(1 - \beta/2)E + \rho_b(t). \tag{26.5}$$

If we substitute $\rho(t)$ into the equation for the density matrix (16.5), we obtain the equation for matrix $\rho_b(t)$,

$$i\hbar\dot{\rho}_b(t) = [\mathcal{H}(t), \rho_b(t)]. \tag{26.6}$$

Note, that this equation has the same form as the equation for the density matrix, $\rho(t)$. If we substitute (26.5) into (16.19) and put $t = 0$, we get the initial condition (26.4) for the matrix $\rho_b(t)$.

Now we consider, for comparison, the "pure" quantum ensemble of noninteracting spins at zero temperature. If the initial state of the ensemble is the ground state, we can describe the evolution of each spin either by the Schrödinger equation, or by the equation for the density matrix (16.5) with the initial conditions,

$$\rho(0) = \begin{pmatrix} 1 & 0 \\ 0 & 0 \end{pmatrix}. \tag{26.7}$$

What is the difference between the average spin for a pure quantum ensemble at zero temperature, and the same ensemble at high temperature?
They differ only by the factor $\beta/2$ which appears in the initial condition
(26.4). The evolution of these two systems will be identical. We already
discussed this result in Chapter 16 for the particular case of a one-qubit
rotation.

To utilize this conclusion for quantum logic gates, we shall try to
obtain a similar conclusion for an ensemble of spin groups (molecules).
Let us consider the simplest case of two interacting spins, with the
Hamiltonian (22.2), and the energy levels shown in Fig. 22.1. In decimal
notation,

$$|00\rangle \to |0\rangle, \quad |01\rangle \to |1\rangle, \quad |10\rangle \to |2\rangle, \quad |11\rangle \to |3\rangle, \quad (26.8)$$

the density matrix has the components, ρ_{ik}, $i, k = 0, 1, 2, 3$. The equilibrium density matrix is given by the expression,

$$\rho_{kk} = \frac{e^{-E_k/k_B T}}{\sum_{k=0}^{3} e^{-E_k/k_B T}}, \quad (\rho_{ik} = 0, \quad i \neq k),$$

where E_k are the energy levels shown in Fig. 22.1. Taking into consideration the inequality, $E_k/k_B T \ll 1$, the density matrix can be approximately represented as,

$$\rho = \frac{1}{4}E + \frac{\hbar}{8k_B T} \times \quad (26.9)$$

$$\begin{pmatrix} \omega_1 + \omega_2 + J & 0 & 0 & 0 \\ 0 & \omega_1 - \omega_2 - J & 0 & 0 \\ 0 & 0 & -\omega_1 + \omega_2 - J & 0 \\ 0 & 0 & 0 & -\omega_1 - \omega_2 + J \end{pmatrix}.$$

Next, we shall assume, for simplicity, that,

$$(\omega_1 - \omega_2), J \ll (\omega_1 + \omega_2)/2. \quad (26.10)$$

In this approximation, we can ignore the influence of the frequency difference, $(\omega_1 - \omega_2)$, and the interaction constant, J, on the population

of the states in the thermal equilibrium. Then, the second term in (26.9) can be rewritten in the simple form,

$$
\rho_\Delta = \frac{\beta}{4}
\begin{pmatrix}
1 & 0 & 0 & 0 \\
0 & 0 & 0 & 0 \\
0 & 0 & 0 & 0 \\
0 & 0 & 0 & -1
\end{pmatrix},
\qquad (26.11a)
$$

where a new β is introduced, $\beta = \hbar(\omega_1 + \omega_2)/2k_B T$. The matrix ρ_Δ, in turn, can be written as a sum of two matrices, ρ_a and ρ_b,

$$
\rho_a = -\frac{\beta}{4}
\begin{pmatrix}
0 & 0 & 0 & 0 \\
0 & 0 & 0 & 0 \\
0 & 0 & 0 & 0 \\
0 & 0 & 0 & 1
\end{pmatrix},
\qquad (26.11b)
$$

$$
\rho_b = \frac{\beta}{4}
\begin{pmatrix}
1 & 0 & 0 & 0 \\
0 & 0 & 0 & 0 \\
0 & 0 & 0 & 0 \\
0 & 0 & 0 & 0
\end{pmatrix}.
$$

To clarify the sense of the representation (26.11b), let us rewrite the matrices ρ_a and ρ_b in binary notation,

$$
\rho_a = -(\beta/4)|11\rangle\langle11|, \qquad \rho_b = (\beta/4)|00\rangle\langle00|. \qquad (26.12)
$$

One can see, that with the accuracy up to constant, $\pm\beta$, the matrix ρ_b describes a pure quantum state, $|00\rangle$, and the matrix ρ_a describes the state, $|11\rangle$.

The question is if we can consider the evolution of the sub-ensemble which can be described, for example, by the density matrix ρ_b with the initial condition, $(\beta/4)|00\rangle\langle00|$, independently of the evolution of the other sub-ensemble, described by the matrix ρ_a. (Of course, as usual, we consider the evolution during a time interval which is short in comparison with the relaxation time.) To answer this question, let us suppose that we apply electromagnetic pulses with the frequency, $(\omega_2 + J)$, which correspond to the resonant transitions between the states $|00\rangle$ and

$|01\rangle$. In this case, we manipulate only the sub-ensemble "b", which can be described by the matrix $\rho_b(t)$,

$$\rho_b(t) = \begin{pmatrix} \rho_{00} & \rho_{01} & 0 & 0 \\ \rho_{10} & \rho_{11} & 0 & 0 \\ 0 & 0 & 0 & 0 \\ 0 & 0 & 0 & 0 \end{pmatrix}. \tag{26.13}$$

The matrix (26.13) satisfies the equation of motion similar to (26.6) with the initial conditions (26.11b). In other words, the equations of motion for the matrix elements, ρ_{ik} ($i, k = 0$ or 1), and $\rho_{n,m}$ ($n, m = 2$ or 3) are approximately decoupled, i.e. the coupled terms in the equations (26.6) are not significant for the evolution of the system. So, again we can manipulate with the "sub-ensemble", "b", which, with the accuracy to the constant, can evolve like an ensemble of "pure" two-level quantum systems, which are initially populated in the ground state. (Certainly, in the same way we could manipulate with the other "sub-ensemble", "a", which is, with accuracy to the constant, equivalent to an ensemble of "pure" two-level systems initially populated in the excited state.)

Finally, if we apply the electromagnetic pulses with the frequency $(\omega_2 + J)$, we obtain the evolution of a two-spin ensemble at room temperatures, which is approximately equivalent to the evolution of an ensemble of noninteracting spins, $I = 1/2$, that are initially populated in the ground state. Roughly speaking, if we only manipulate with the right spin, and do not touch the left spin, the evolution of the right spin at room temperature will be the same as the evolution of a single spin, $I = 1/2$, which initially is populated in the ground state.

However, we should note that we did not achieve our main purpose – to realize quantum logic gates at high temperature. In fact, an ensemble of two-spin systems does not have any advantage over the ensemble of one-spin systems. Let us take the next step. We shall consider an ensemble of non-interacting four-spin molecules, for example,

$$\begin{array}{|cc|} \hline A & B \\ C & D \\ \hline \end{array},$$

in which all spins are connected by the Ising interaction. We assume, for simplicity, that the difference between the one-spin transition frequencies is small compared with the value of the average frequency, $\sum \omega_k / 4$, so we can ignore the influence of these differences on the population of the levels. Then, the equilibrium density matrix is given by the expression,

$$\rho = \frac{1}{16} E + \rho_\Delta, \qquad (26.14)$$

where the deviation matrix, ρ_Δ, has the decimal notation form,

$$\rho_\Delta = \frac{\beta}{16} \{2|0\rangle \langle 0| + |1\rangle \langle 1| + |2\rangle \langle 2| + |4\rangle \langle 4| + |8\rangle \langle 8| - \qquad (26.15)$$

$$(|7\rangle \langle 7| + |11\rangle \langle 11| + |13\rangle \langle 13| + |14\rangle \langle 14|) - 2|15\rangle \langle 15|\},$$

$$(\beta = \sum_{k=0}^{3} \hbar \omega_k / 4k_B T).$$

In binary notation the deviation matrix (26.15) can be written as,

$$\rho_\Delta = \frac{\beta}{16} \{2|0000\rangle \langle 0000| + |0001\rangle \langle 0001| + \qquad (26.16)$$

$$|0010\rangle \langle 0010| + |0100\rangle \langle 0100| + |1000\rangle \langle 1000| -$$

$$(|0111\rangle \langle 0111| + |1011\rangle \langle 1011| + |1101\rangle \langle 1101| +$$

$$+|1110\rangle \langle 1110|) - 2|1111\rangle \langle 1111|\},$$

$$|ijnm\rangle \langle ijnm| = |i_A j_B n_C m_D\rangle \langle i_A j_B n_C m_D|.$$

Using a unitary transformation, we would like to transform (26.16), to get the desirable sub-ensembles. We shall discuss later, in Chapter 28, how to make such transformations. Now let us rewrite the diagonal deviation matrix (26.16) in the form of the table (see Tbl. 26.1). In Tbl. 26.1, the first row shows all possible matrices, $|ijnm\rangle \langle ijnm|$, which are denoted by a block,

i	j
n	m

00	00	00	00
00	01	10	11
2	1	1	0
2	0	0	0

01	01	01	01	10	10	10	10	11	11	11	11
00	01	10	11	00	01	10	11	00	01	10	11
1	0	0	-1	1	0	0	-1	0	-1	-1	-2
-1	1	1	1	1	-1	-1	-1	-2	0	0	0

Table 26.1: The diagonal matrix elements of ρ_Δ before (the second row) and after (the third row) redistribution. The upper part of the table represents the transformation to the ground state of the effective two-spin system. The lower table represents the transformation of the remaining diagonal elements of the density matrix.

The second row shows the coefficients for the corresponding matrices in (26.16), without the factor $\beta/16$. Assume, using a unitary transformation, we manage to redistribute the coefficients in (26.16) as shown in the third row of Tbl. 26.1. Consider the first four columns in Tbl. 26.1. They correspond to the states, $|00ij\rangle$. If we apply to our molecule $ABCD$, the electromagnetic pulses which induce transitions between the states $|00ij\rangle$, we will never disturb spins A and B. We can consider a sub-ensemble of these states, and they evolve approximately as a "pure" two-spin system at zero temperature, which is initially populated in the ground state. We shall discuss this case in the next chapter.

Chapter 27

Evolution of an Ensemble of Four-Spin Molecules

To explain the dynamics of an ensemble of four-spin molecules, let us write the Hamiltonian of the complex,

A	B
C	D

taking into consideration the Ising interaction between all spins [64]. To get the Hamiltonian in a system of coordinates which rotates with the frequency of the circular polarized magnetic field, ω, we use the unitary transformation (15.5a),

$$\Psi' = e^{-i\omega I^z t}\Psi, \tag{27.1}$$

$$\mathcal{H}' = e^{-i\omega I^z t}\mathcal{H}e^{i\omega I^z t},$$

$$I^z = I_0^z + I_1^z + I_2^z + I_3^z,$$

where the lower index, "0", indicates the last spin, D; index "1" refers to the spin C, and so on. Using the method, which was described in Chapter 15, we get the time-independent Hamiltonian in the rotating frame,

$$\mathcal{H}' = -\hbar \left\{ \sum_{k=0}^{3} (\omega_k - \omega) I_k^z + 2 \sum_{i,k=0}^{3} J_{ik} I_i^z I_k^z + \sum_{k=0}^{3} \Omega_k I_k^x \right\}, \quad (27.2)$$

where, in the second term we suppose $i < k$, $\omega_k = \gamma_k B^z$, $\Omega_k = \gamma_k h$. When the rotating magnetic field is absent ($\Omega_k = 0$, $\omega = 0$), the Hamiltonian (27.2) has only diagonal matrix elements, which determine the energies of the stationary states. To find these energies, we should rewrite the expression for I_m^z (12.4),

$$I_m^z = \frac{1}{2}(|0_m\rangle\langle 0_m| - |1_m\rangle\langle 1_m|), \quad (m = 0, 1, 2, 3),$$

in terms of the four-spin basic states, $|i_3 j_2 k_1 n_0\rangle$. So, we have,

$$I_0^z = \frac{1}{2}(|i_3 j_2 k_1 0_0\rangle\langle i_3 j_2 k_1 0_0| - |i_3 j_2 k_1 1_0\rangle\langle i_3 j_2 k_1 1_0|), \quad (27.3)$$

$$I_1^z = \frac{1}{2}(|i_3 j_2 0_1 n_0\rangle\langle i_3 j_2 0_1 n_0| - |i_3 j_2 1_1 n_0\rangle\langle i_3 j_2 1_1 n_0|),$$

$$I_2^z = \frac{1}{2}(|i_3 0_2 k_1 n_0\rangle\langle i_3 0_2 k_1 n_0| - |i_3 1_2 k_1 n_0\rangle\langle i_3 1_2 k_1 n_0|),$$

$$I_3^z = \frac{1}{2}(|0_3 j_2 k_1 n_0\rangle\langle 0_3 j_2 k_1 n_0| - |1_3 j_2 k_1 n_0\rangle\langle 1_3 j_2 k_1 n_0|),$$

where we assume summation over the indices i_3, j_2, k, and n_0. In the same way, we can write the second term in the Hamiltonian (27.2) in terms of the four-spin basic states, $|i_3 j_2 k_1 n_0\rangle$,

$$I_0^z I_1^z = \frac{1}{2}(|0_0\rangle\langle 0_0| - |1_0\rangle\langle 1_0|) \cdot \frac{1}{2}(|0_1\rangle\langle 0_1| - |1_1\rangle\langle 1_1|) \rightarrow \quad (27.4)$$

$$\frac{1}{4}(|i_3 j_2 0_1 0_0\rangle\langle i_3 j_2 0_1 0_0| + |i_3 j_2 1_1 1_0\rangle\langle i_3 j_2 1_1 1_0| -$$

$$|i_3 j_2 0_1 1_0\rangle\langle i_3 j_2 0_1 1_0| - |i_3 j_2 1_1 0_0\rangle\langle i_3 j_2 1_1 0_0|).$$

Analogous expressions can be written for the other terms, $I_i^z I_j^z$. Next, we omit indices in the expressions for four-basic states.

Now, we can find the energies of the stationary states. For the state $|0000\rangle$ we have the energy,

$$E = E_0 = -\frac{\hbar}{2}\left(\sum_{k=0}^{3}\omega_k + \sum_{i,k=0}^{3}J_{ik}\right), \quad (i < k). \tag{27.5}$$

For the state, $|0001\rangle$, we get,

$$E = E_0 + \hbar\left(\omega_0 + \sum_{k=1}^{3}J_{0k}\right). \tag{27.6}$$

For the states $|0010\rangle$, $|0100\rangle$, and $|1000\rangle$ we have, correspondingly,

$$E = E_0 + \hbar\left(\omega_1 + \sum_{k=0}^{3}J_{1k}\right), \quad (k \neq 1), \tag{27.7}$$

$$E = E_0 + \hbar\left(\omega_2 + \sum_{k=0}^{3}J_{2k}\right), \quad (k \neq 2),$$

$$E = E_0 + \hbar\left(\omega_3 + \sum_{k=0}^{2}J_{3k}\right).$$

In (27.7) we assume that $J_{ik} = J_{ki}$. The first five energy levels are shown in Fig. 27.1, where we suppose that $\omega_n < \omega_{n+1}$.

The Hamiltonian of our system (27.2) includes also the nondiagonal elements, which appear due to the rotating magnetic field (the third term in (27.2)). To derive the nondiagonal elements, we should write the operator I_m^x (12.10),

$$I_m^x = \frac{1}{2}(|0_m\rangle\langle 1_m| + |1_m\rangle\langle 0_m|),$$

in terms of basic states, $|ijkn\rangle$, in the same way as we did for the diagonal elements. We have,

$$I_0^x = \frac{1}{2}(|ijk0\rangle\langle ijk1| + |ijk1\rangle\langle ijk0|), \tag{27.8}$$

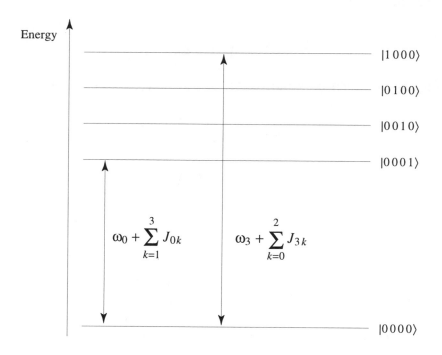

Figure 27.1: The first five energy levels for the four-spin system.

$$I_1^x = \frac{1}{2}(|ij0n\rangle\langle ij1n| + |ij1n\rangle\langle ij0n|),$$

$$I_2^x = \frac{1}{2}(|i0kn\rangle\langle i1kn| + |i1kn\rangle\langle i0kn|),$$

$$I_3^x = \frac{1}{2}(|0jkn\rangle\langle 1jkn| + |1jkn\rangle\langle 0jkn|).$$

Finally, we transform to decimal notation,

$$|0000\rangle \to |0\rangle, \quad |0001\rangle \to |1\rangle, \quad |0010\rangle \to |2\rangle,$$

and so on. Then, combining the coefficients of $|0000\rangle\langle 0000|$ in the Hamiltonian (27.2), we get the matrix element, $\mathcal{H}_{00} = E_0$, which is given by the expression (27.5). The matrix element, $\mathcal{H}_{11}$ is given by (27.6). The matrix elements, $\mathcal{H}_{22}$, $\mathcal{H}_{44}$, and $\mathcal{H}_{88}$ are given by (27.7), and so on. In the same way, combining the coefficients at $|0000\rangle\langle 0001|$, we get the nondiagonal matrix element,

$$\mathcal{H}_{01} = -\frac{\hbar}{2}\Omega_0. \tag{27.9}$$

Analogously, we can get other nondiagonal matrix elements. For example, for even n, $n \leq 14$, we have, $\mathcal{H}_{n,n+1} = \mathcal{H}_{01}$. For all k, $0 \leq k \leq 7$, we get,

$$\mathcal{H}_{k,k+8} = -\frac{\hbar}{2}\Omega_3, \tag{27.10}$$

and so on. Because the Hamiltonian is a Hermitian matrix, for real nondiagonal elements we have, $\mathcal{H}_{ik} = \mathcal{H}_{ki}$.

Now we are ready to write the equations of motion for the matrix elements of the density matrix. We put,

$$\rho = \rho_\infty + (\beta/16)r, \quad \beta = \hbar \sum_{k=0}^{3} \omega_k/4k_BT, \tag{27.11}$$

where, as before, $\rho_\infty = E/16$ is the density matrix which corresponds to the infinite temperature. From (16.5) we have the following equations for the matrix elements r_{ik},

$$i\hbar\dot{r}_{ik} = \mathcal{H}_{in}r_{nk} - r_{in}\mathcal{H}_{nk}, \quad 0 \leq i, k \leq 15. \tag{27.12}$$

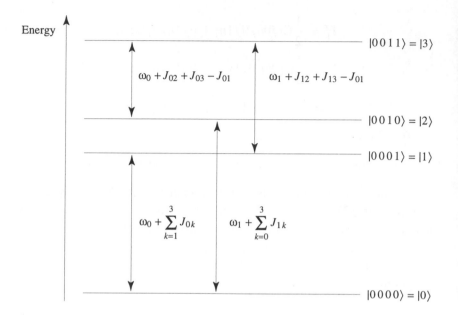

Figure 27.2: Energy levels and frequencies of one-spin transitions for the states $|00ij\rangle$.

We shall present here the explicit equations for r_{00} and r_{01},

$$-2i\dot{r}_{00} = \Omega_0(r_{10} - r_{01}) + \Omega_1(r_{20} - r_{02}) + \qquad (27.13)$$

$$\Omega_2(r_{40} - r_{04}) + \Omega_3(r_{80} - r_{08}),$$

$$-2i\dot{r}_{01} = (2/\hbar)(E_1 - E_0)r_{01} + \Omega_0(r_{11} - r_{00}) + \Omega_1(r_{21} - r_{03}) +$$

$$\Omega_2(r_{41} - r_{05}) + \Omega_3(r_{81} - r_{09}),$$

where E_1 is the energy of the state $|0001\rangle$.

Now, let us assume that we are going to induce single spin transitions only between the states, $|00ij\rangle$. The corresponding energy levels and frequencies are shown in Fig. 27.2. If we apply the electromagnetic pulses with the frequencies shown in Fig. 27.2, we can induce only the transitions shown in this figure. This means that the only matrix elements of the density matrix (or matrix r) which undergo a significant change are the following,

$$r_{00}, \quad r_{11}, \quad r_{22}, \quad r_{33}, \tag{27.14}$$

$$r_{01}, \quad r_{02}, \quad r_{23}, \quad r_{13},$$

$$r_{10}, \quad r_{20}, \quad r_{32}, \quad r_{31}.$$

The nondiagonal matrix elements, r_{01} and r_{10} correspond to a change of the orientation of spin D, when spin C is in the ground state. The matrix elements, r_{02} and r_{20} describe the inversion of spin C while spin D is in the ground state, and so on. All four transitions correspond to the ground state of spins A and B.

Now we consider the equations (27.13). For our case, neglecting all nondiagonal terms, except (27.14), we can rewrite the equations (27.13) as,

$$-2i\dot{r}_{00} = \Omega_0(r_{10} - r_{01}) + \Omega_1(r_{20} - r_{02}), \tag{27.15}$$

$$-2i\dot{r}_{01} = (2/h)(E_1 - E_0)r_{01} + \Omega_0(r_{11} - r_{00}).$$

Similar approximate equations can be written for all matrix elements (27.14). Because of the decoupling from the other matrix elements, these equations do not differ from the corresponding equations for two spin system.

Now, let us assume that the initial density matrix of the four-spin system has diagonal matrix elements corresponding to the third row in Tbl. 26.1. Then, for the matrix elements (27.14), the initial conditions are given by the expressions,

$$r_{00}(0) = 2, \quad r_{ik}(0) = 0, \quad (i, k = 0, 1, 2, 3), \quad (i, k) \neq (0, 0).$$
$$\tag{27.16}$$

The approximate equations for the matrix elements (27.14), in combination with the initial conditions (27.15), do not differ (with the accuracy to the constant) from the case when one has a "pure" two-spin system, at zero temperature, which is initially populated in the ground state. So, quantum logic can be realized using two-spin sub-ensemble of the four-spin system, at room temperature, if one can transform the initial density matrix in such a way that the matrix, r, will have the first matrix elements, r_{ik}, $(i, k = 0, 1, 2, 3)$ given by (27.16).

Chapter 28

Getting the Desired Density Matrix

Now we shall discuss how to make the transformation,

$$a_{iknm}|iknm\rangle\langle iknm| \rightarrow a'_{iknm}|iknm\rangle\langle iknm|, \qquad (28.1)$$

where we assume summation over the indices i, k, n, and m; a_{iknm} are the numbers in the second row of Tbl. 26.1; a'_{iknm} are the numbers in the third row of Tbl. 26.1. Ignoring the factor, $\beta/16$, these numbers represent the diagonal elements of the deviation matrix. To realize this transformation, Gershenfeld and Chuang [29] applied the next sequence (GC-sequence) of the CN-gates to the initial density matrix (26.16),

$$GC = \text{CN}_{02}\text{CN}_{12}\text{CN}_{21}\text{CN}_{20}. \qquad (28.2)$$

To check the action of this sequence, let us first consider the transformation of the density matrix under the action of the unitary operator, U. Consider the wave function,

$$\Psi = \sum c_n |n\rangle. \qquad (28.3)$$

After the action of the unitary operator, U, we get new wave function,

$$\Psi' = \sum c'_n |n\rangle, \qquad (28.4)$$

where the new coefficients, c'_n, are expressed in terms of the old ones, c_n, in the following way,

$$c'_n = U_{nk} c_k, \qquad (28.5)$$

where we assume summation over repeated indices. The matrix elements of the new density matrix, after the unitary transformation U, can be represented as,

$$\rho'_{nm} = c'_n c'_m{}^* = U_{nk} c_k U^*_{mp} c^*_p = \qquad (28.6)$$

$$U_{nk} U^\dagger_{pm} c_k c^*_p = U_{nk} \rho_{kp} U^\dagger_{pm},$$

where we used the notation,

$$\rho_{nm} = c_n c^*_m, \qquad U^\dagger_{pm} = U^*_{mp}. \qquad (28.7)$$

Formula (28.6) represents the well-known quantum-mechanical equation for the transformation of the density matrix,

$$\rho' = U \rho U^\dagger. \qquad (28.8)$$

In our case, $U = GC$ (28.2).

Let us check, for example, the action of the GC-sequence on the last term in (26.16). Up to a factor, $-\beta/8$, we have the initial matrix, $M_0 = |1_3 1_2 1_1 1_0\rangle\langle 1_3 1_2 1_1 1_0|$. Now, we find the transformation of this matrix under the action of the first gate CN$_{20}$ in (28.2),

$$\text{CN}_{20} = |0_2 0_0\rangle\langle 0_2 0_0| + |0_2 1_0\rangle\langle 0_2 1_0| \ | \qquad (28.9a)$$

$$|1_2 1_0\rangle\langle 1_2 0_0| + |1_2 0_0\rangle\langle 1_2 1_0|, \quad \text{CN}^\dagger_{20} = \text{CN}_{20}.$$

We have,

$$M'_1 = \text{CN}_{20} M_0 = \text{CN}_{20} \cdot |1_3 1_2 1_1 1_0\rangle\langle 1_3 1_2 1_1 1_0| = \qquad (28.9b)$$

$$|1_3 1_2 1_1 0_0\rangle\langle 1_3 1_2 1_1 1_0|,$$

$$M_1 = M'_1 \text{CN}^\dagger_{20} = |1_3 1_2 1_1 0_0\rangle\langle 1_3 1_2 1_1 0_0|.$$

One can see, that the action of CN_{20} on the matrix, $|i_3 j_2 n_1 k_0\rangle \langle i_3 j_2 n_1 k_0|$, is the following. If $j = 0$, then, the matrix does not change; if $j = 1$, then,

$$|ijnk\rangle\langle ijnk| \rightarrow |ijn\bar{k}\rangle\langle ijn\bar{k}|, \tag{28.10}$$

where $\bar{k}$ means "complement" to k ($\bar{0} = 1$, $\bar{1} = 0$). So, the action of the CN-gate on the density matrix is similar to its action on the quantum state. After application of the CN_{21}-gate, we have,

$$M_2 = CN_{21} M_1 CN_{21}^{\dagger} = |1_3 1_2 0_1 0_0\rangle\langle 1_3 1_2 0_1 0_0|. \tag{28.11}$$

After application of CN_{12} and CN_{02}, the matrix M_2 does not change, as in this case, the control units are $|0_1\rangle\langle 0_1|$ and $|0_0\rangle\langle 0_0|$, correspondingly. So, the matrix $|1_3 1_2 1_1 1_0\rangle\langle 1_3 1_2 1_1 1_0|$ transforms to the matrix, $|1_3 1_2 0_1 0_0\rangle\langle 1_3 1_2 0_1 0_0|$. This corresponds to Tbl. 26.1 (compare the last column of the second row, and the 4-th to the end column of the third row). In the same way, considering, for example, the second term in (26.16), we obtain the following transformation under the action of the GC sequence (28.2),

$$0. \quad M_0 = |0_3 0_2 0_1 1_0\rangle\langle 0_3 0_2 0_1 1_0|, \tag{28.12}$$

$$1. \quad M_1 = CN_{20} M_0 CN_{20}^{\dagger} = M_0,$$

$$2. \quad M_2 = CN_{21} M_0 CN_{21}^{\dagger} = M_0,$$

$$3. \quad M_3 = CN_{12} M_0 CN_{12}^{\dagger} = M_0,$$

$$4. \quad M_4 = CN_{02} M_0 CN_{02}^{\dagger} = |0_3 1_2 0_1 1_0\rangle\langle 0_3 1_2 0_1 1_0|.$$

This transformation also corresponds to Tbl. 26.1. In the same way, we can check all other transformations.

Thus, using the GC sequence of the CN-gates, one can transform the initial density matrix to the density matrix which describes a subensemble of spins in the states $|0_3 0_2 i_1 j_0\rangle$, whose evolution corresponds to the evolution of an ensemble of two-spin systems, which are initially populated in the ground state. Analogously, one can get the effective 3-spin "pure" quantum system from a 6-spin chain, and so on [29]. Two-spin effective system can be used for implementation of the two-qubit gates. Bigger effective systems will be, probably, convenient for the quantum computation.

Currently, such experiments are expected to be done with many-atomic molecules in liquids. A big molecule can involve a number of weakly interacting nuclear spins (usually protons), which have a slightly different frequencies depending on the chemical structure. The interaction between molecules is very small, and the times of relaxation are extremely large: the smallest time which corresponds to the relaxation of the transversal component of the average nuclear spin is of the order of 1s. Because of very small differences between the frequencies of spins, the special complicated sequences of pulses are expected to be used to manipulate with a spin. So, one can not exclude that the first quantum computation will be done not in a powerful ion trap, but, as it was mentioned by Gary Taubes [31], in a cup of coffee.

Chapter 29

Conclusion

Here we present our vision of the current stage of quantum computation. We mentioned in the Introduction that there exist two main directions for design of future computers. One of them is connected with the development of digital computers, and is based on electron conductivity. The other direction—of quantum computation—is connected with development of quantum computers, and is based mainly on the resonant interaction of electromagnetic pulses with nuclear or atomic systems. The output of quantum computation, in a simple variant, is a sequence of data, "there is voltage" (which represents "1"), and "there is no voltage" (which represents "0"). There exist other suggestions for implementations of quantum computation, for example, those using the spin states of coupled single-electron quantum dots [65]. These systems do not use resonance pulses, and could be of significant interest for quantum computation.

The problem of decoherence was not addressed in this book. It is the main obstacle for the physical realization of a quantum computer. An entangled pair of qubits is a superposition of two qubits that cannot be decomposed. In a closed system it would remain in that superposition indefinitely. But no system is closed and the interactions with the environment destroy this delicate state. The pair of qubits has decohered. Initial estimates of the decoherence time [66] were not encouraging, but new systems seem to offer longer decoherence times [38]. Quantum computation may inspire new directions in material science, as we search for materials that have long decoherence times.

When qubits are in superposition, they are not in an eigenstate of the Hamiltonian describing the quantum computer. Their dynamics then becomes important. This establishes a relation between dynamical systems and quantum computation. Most of the models used for quantum computation are quantum chaotic systems. This means that treated as a classical system they are chaotic. We feel that for the physical realization of a quantum computer, the dynamical processes should be understood. Recent developments in dynamical systems have provided us with the tools that allow us to explore this new field of dynamical quantum computation.

So far, there are several significant achievments in quantum computing: the first quantum algorithm, the first error correction codes, and two very promising implementations of quantum logic (the cooled ions in ion traps, and nuclear spins in molecules). The most important future step is experimental implementation of quantum logic. A real quantum logic gate should demonstrate the correct transformation of an arbitrary superpositional state, taking into consideration both magnitude and phase of complex amplitudes. It is possible that all difficulties of quantum computation will be overcome. The unexpected discoveries of the last few years make us feel optimistic.

Bibliography

[1] M.A. Kastner, *Rev. Mod. Phys.*, **64**, (1992) 849.

[2] F. A. Buot, *Phys. Rep.*, **234**, (1993) 73.

[3] K. Yano, T. Ishii, T. Hashimoto, T. Kobayashi, F. Murai, K. Seki, *IEEE Transactions on Electron Devices*, **41**, (1994) 1628.

[4] S. Tiwari, F. Rana, H. Hanafi, A. Hartstein, E.F. Crabbé, K. Chan, *Appl. Phys. Lett.*, **68**, (1996) 1377.

[5] D. Goldhaber-Gordon, M.S. Montemerdo, J.C. Love, G.J. Opiteck, J.C. Ellenbogen, *Proceedings of the IEEE*, **85**, (1997) 521.

[6] R.F. Service, *Science*, **275**, (1997) 303.

[7] L. Guo, E. Leobandung, S.Y. Chou, *Science*, **275**, (1997) 649.

[8] A. Aviram, *Int. J. Quantum Chem.*, **42**, (1992) 1615.

[9] S.V. Subramanyam, *Current Sci.*, **67**. (1994) 844.

[10] M. Dresselhaus, *Phys. World*, **9**, (1996) 18.

[11] L. Kouwenhoven, *Science*, **275**, (1997) 1896.

[12] R. Landauer, *IBM J. Res. Develop.*, **5**, (1961) 183.

[13] C.H. Bennett, *IBM J. Res. Develop.*, **17**, (1973) 525.

[14] T. Toffoli, In: *Automata, Languages and Programming*, Eds J.W. de Bakker, J. van Leeuwen, Springer-New York, (1980), p. 632.

[15] P. Benioff, *J. Stat. Phys.*, **22**, (1980) 563; *J. Stat. Phys.*, **29**, (1982) 515.

[16] R. P. Feynman, *Int. Journal of Theor. Phys.*, **21**, (1982) 467; *Opt. News*, **11**, (1985) 11.

[17] D. Deutsch, *Proc. R. Soc. London, Ser. A*, **425**, (1989) 73.

[18] S. Lloyd, *Science*, **261**, (1993) 1569.

[19] P. Shor, *Proc. of the 35th Annual Symposium on the Foundations of Computer Science, IEEE, Computer Society Press*, New York, (1994), p. 124.

[20] A. Barenco, C.H. Bennett, R. Cleve, D.P. DiVincenzo, N. Margolus, P. Shor, T. Sleator, J. Smolin, H. Weinfurter, *Phy. Rev. A*, **52**, (1995) 3457.

[21] J.I. Cirac, P. Zoller, *Phys. Rev. Lett.*, **74**, (1995) 4091.

[22] C. Monroe, D.M. Meekhof, B.E. King, W.M. Itano, D.J. Wineland, *Phys. Rev. Lett.*, **75**, (1995) 4714.

[23] R.J. Hughes, D.F. James, J.J. Gomez, M.S. Gulley, M.H. Holzscheiter, P.G. Kwiat, S.K. Lamoreaux, C.G. Peterson, V.D. Sandberg, M.M. Schauer, C.E. Thorburn, D. Tupa, P.Z. Wang, A.G. White, LA-UR-97-3301, quant-ph/9708050.

[24] Q.A. Turchette, C.J. Hood, W. Lange, H. Mabuchi, H.J. Kimble, *Phys. Rev. Lett.*, **75**, (1995) 4710.

[25] P.W. Shor, *Phys. Rev. A*, **52**, (1995) R2493.

[26] L.K. Grover, *Proceedings, STOC*, 1996, pp.212-219.

[27] L.K. Grover, *Phys. Rev. Lett.,* **79**, (1997) 325.
G. Brassard, *Science*, **275**, (1997) 627.
G.P. Collins, *Physics Today*, October, (1997) 19.

[28] N. Gershenfeld, I. Chuang, S. Lloyd, *Phys. Comp. 96, Proc. of the Fourth Workshop on Physics and Computation*, 1996, p. 134.

[29] N.A. Gershenfeld, I.L. Chuang, *Science*, **275**, (1997) 350.

[30] D.G. Cory, A.F. Fahmy, T.F. Havel, *Phys. Comp. 96, Proc. of the Fourth Workshop on Physics and Computation*, 1996, p. 87; *Proc. Natl. Acad. Sci. USA*, **94**, (1997) 1634.

[31] G. Taubes, *Science*, **275**, (1997) 307.

[32] R. Laflamme, E. Knill, W.H. Zurek, P. Catasti, S. Velupillai, quant-ph/9769025.

[33] I.L. Chuang, R. Laflamme, P.W. Shor, W.H. Zurek, *Science*, **270**, (1995) 1633.

[34] W.H. Zurek, J.P. Paz, *Il Nuovo Cimento*, **110B**, (1995) 611.

[35] S. Lloyd, *Scientific American*, **273**, (1995) 140.

[36] C.H. Bennett, *Physics Today*, **48**, October (1995) 24.

[37] C.H. Bennett, D.P. DiVincenzo, *Nature*, **377**, (1995) 389.

[38] D.P. DiVincenzo, *Science*, **270**, (1995) 255.

[39] A.E. Ekert, R. Jozsa, *Rev. Mod. Phys.*, **68**, (1996) 733.

[40] C. Williams and S. Clearwater, *Explorations in Quantum Computing*, Springer-Verlag, 1997.

[41] D. DiVincenzo, E. Knill, R. Laflamme, and W. Zurek, eds., Proceedings of the ITP Conference on Quantum Coherence and Decoherence, published in *Proc. of the Royal Society of London*, **454** (1998) 257–486.

[42] D. Deutsch, Artur Ekert, *Physics World*,**11** (1998) 47.

[43] D. DiVincenzo, B. Terhal, *Physics World*,**11** (1998) 53.

[44] A.M. Turing, *Proc. Lond. Math. Soc., Sec. 2*, **43**, (1936) 544.

[45] I. Adler, *Thinking Machines*, The John Day Company, New York, 1974.

[46] R.P. Feynman, *Feynman Lectures on Computation,*   Addison-Wesley Publishing Company, 1996.

[47] E. Fredkin, T. Toffoli, *Inter. J. Theor. Phys.*, **21**, (1982) 219.

[48] J.J. Sakuraii, *Modern Quantum Mechanics*, Addison-Wesley Publishing Company, 1995.

[49] W. Paul, *Rev. Mod. Phys.*, **62**, (1990) 531.

[50] M.G. Raizen, J.M. Gilligan, J.C. Bergquist, W.M. Itano, D.J. Wineland, *Phys. Rev. A*, **45**, (1992) 6493.

[51] H. Walther, *Adv. in At. Mol. and Opt. Phys.*, **32**, (1994) 379.

[52] G.P. Berman, G.D. Doolen, G.D. Holm, V.I. Tsifrinovich, *Phys. Lett. A*, **193**, (1994) 444.

[53] A. Barenco, D. Deutsch, A. Ekert, R. Jozsa, *Phys. Rev. Lett.*, **74**, (1995) 4083.

[54] G.P. Berman, D.K. Campbell, G.D. Doolen, G.V. López, V.I. Tsifrinovich, *Physica B*, **240**, (1997) 61.

[55] G.P. Berman, D.K. Campbell, V.I. Tsifrinovich, *Phys. Rev. B*, **55**, (1997) 5929.

[56] K.N. Alckseev, G.P. Berman, V.I. Tsifrinovich, A.M. Fishman, *Sov. Phys. Usp.*, **35**, (1992) 572.

[57] T. Sleator, H. Weinfurter, *Phys. Rev. Lett.*, **74**, (1995) 4087.

[58] *Cavity Quantum Electrodynamics*, Ed. P.A. Berman, *Academic Press*, New York, 1994.

[59] N.F. Ramsey, *Molecular Beams, Oxford University Press*, 1985.

[60] R. Landauer, *Philos. Trans. R. Soc. London*, **353**, (1995) 367.

[61] A. Steane, *Proc. R. Soc. London, Se. A*, **452**, (1996) 2551.

[62] R. Laflamme, C. Miquel, J.P. Paz, W.H. Zurek, *Phys. Rev. Lett.,* **77**, (1996) 198.

[63] G.P. Berman, G.D. Doolen, G.V. López, V.I. Tsifrinovich, quant-ph/9802013.

[64] G.P. Berman, G.D. Doolen, G.V. López, V.I. Tsifrinovich, quant-ph/9802016.

[65] D. Loss, D.P. DiVincenzo, *Phys. Rev. A* **57**, (1998), 120. Also cond-mat/9701055, 1997.

[66] W. G. Unruh, *Phys. Rev. A*, **51** (1995) 992.

Index